CONSCIOUS
VALUE

CONSCIOUS **VALUE**

HOW CONSCIOUSNESS REVEALS WHAT TRULY MATTERS

ADRIAN DAVID NELSON

WAKING COSMOS
2025

Conscious Value: How Consciousness Reveals What Truly Matters

First Printing: 2025
ISBN: 978-1-0369-3454-5
Published by Waking Cosmos
Contact: adrian.david.nelson [at] gmail.com

For Lia and Karam

CONTENTS

PART I. THE REALITY OF CONSCIOUSNESS

1. REAL VALUE 2
2. MORAL OBJECTIVITY 8
3. INTRINSIC AND INSTRUMENTAL VALUE 14
4. HEDONIC VALUE AND IMPARTIALITY 25
5. THE QUALIASCOPE: BEYOND THE HUMAN LENS 39
6. A NEW REALITY 44
7. COSMIC REACH 48

PART II. KNOWN UNKNOWNS

8. INTO THE UNKNOWN 66
9. ANIMAL CONSCIOUSNESS 70
10. ARTIFICIAL MINDS 81
11. VALUE ASYMMETRY 93
12. ALIEN MINDS 106

PART III. BEYOND UTILITY

13. MORAL PROGRESS 124
14. THE ORGAN DONOR 128
15. THE EXPERIENCE MACHINE 133
16. THEORY X 140
17. LIBERATING CONSCIOUSNESS 152
18. THE IDENTITY OF CONSCIOUSNESS 164

APPENDIX. SOLVING THE NON-IDENTITY PROBLEM 180

REFERENCES 194

PART I.

THE REALITY OF CONSCIOUSNESS

1. REAL VALUE

For many years, I've been fascinated by one thing: it is the fact that in our universe, there is this mysterious phenomenon we call "consciousness." Why is it that in our world certain outwardly physical structures also have inner mental states—something *it is like* from the inside?

From mystics and sages, to philosophers and scientists, the mystery of consciousness has captivated thinkers of all kinds for millennia. But beneath this enduring enigma lies a second, often overlooked question. Not just: what is consciousness? But: *why does it matter?* Why is it significant that consciousness exists at all? These two mysteries—one a question of existence, the other a question of value—are intimately tied. To understand the deeper question of why consciousness matters, we must first grapple with the mysterious fact of its existence.

Science has given us profound insight into the physical world—its structure, laws, and patterns—and this insight has transformed the world and human life. And yet the phenomenon of consciousness still resists our best attempts to explain it.

We have, of course, learned a great deal about the brain and nervous system. We have mapped the neural correlates of behaviors, language, memory, and decision-making. And yet while this catalogue of correlations has proven extremely valuable, the central mystery of consciousness remains: Exactly why do certain physical processes produce subjective experience? What is it about these processes that allows them to perform the metaphysical alchemy of creating mind from matter? As the contemporary philosopher Hedda Hassel Mørch puts it:

> "The nature of consciousness seems to be unique among scientific puzzles. Not only do neuroscientists have no fundamental explanation for how it arises from physical states of the brain, we are not even sure whether we ever will." [1]

However mysterious its origin, consciousness seems to be indisputably real. As the 17th-century philosopher René Descartes famously observed, we can question many aspects of our experience, and we are often deceived by our senses. And yet while there is always the possibility that the world we perceive is the product of a dream or a hallucination, the fact that we are right now having an experience is beyond doubt.[2]

Some philosophers have nevertheless attempted to deny the reality of consciousness altogether. Illusionists, such as Daniel Dennett, have argued that subjective experience is merely a cognitive trick: a "user illusion" constructed by the brain with no deeper reality.[3] And yet what has ultimately proven significant about this view is the way in which it fails. As the neuroscientist Sam Harris points out,

> "To say that consciousness may only seem to exist is to admit its existence in full—for if things seem any way at all, that is consciousness." [4]

Today academics continue to debate the nature of consciousness yet most now recognize that whatever it is, it surely does exist. Indeed, it is, in a sense, the most certainly real phenomenon in the universe. It is through it that we become aware of anything else. It is the light by which all other facts are known.

The key question is not just how matter gives rise to mind. If a scientific explanation is ever found, the question would still remain: why is it this process and not some other that transforms lifeless atoms into living, feeling subjectivity? This puzzle is sharpened by the fact that modern science is built to describe reality as it exists from the outside to all observers—in entirely objective and impersonal terms. But consciousness is that which is quintessentially *inside.* It is the private, first-person dimension of reality. The purely objective terms and language of physical science appear inadequate to capture it. This explanatory impasse is known as the hard problem of consciousness, and it has intrigued, captivated, but ultimately eluded all who have so far attempted to resolve it.[5]

The puzzle of consciousness has long been a point of controversy. In Western academic thought, the subject has been largely sidelined, seemingly almost out of embarrassment, for much of the last two centuries. In recent decades, however, the mood has changed. A general shift in attitudes has occurred, and academics and scientists have begun taking a renewed interest in this enigmatic phenomenon and how it might yet be incorporated within our scientific picture of reality.

I've followed this debate for many years with growing fascination. In my own work, I've explored possible solutions to the hard problem of consciousness and what they might imply—not just for consciousness, but for our concepts of the self and even the nature of reality. And yet, while recent years have seen a surge of new interest in this area, I believe that the most important truth about consciousness remains strangely overlooked. That truth is not just that consciousness is a puzzling, even uncanny denizen of the universe, but that it *matters*. It is a domain of real and unmistakable value, both positive and negative. Value which, by extension, gives reality to ethics.

The hard problem of consciousness is a profound mystery, and yet for many it has obscured what is in some ways an even deeper truth—its unmistakable ethical significance. As we will explore, certain conscious states bring us into contact with a real and veridical moral reality: one that transcends history, culture, and species membership. Certain qualities of conscious experience, I will argue, have an objective ethical status that is not dependent on—or created by—our beliefs or judgments about them. In other words, the goodness and badness of certain qualities of consciousness are inherent to them. The clearest example is a negative case: the existence of suffering.

There is, I believe, an undeniable and immutable sense in which it matters that conscious beings can suffer. That all things being equal, suffering is bad, and its potential to occur counts against any action that could lead to it. The phenomenology (or feeling) of suffering—its inherent "ought-not-to-be-ness"—has a force that compels ethical concern, regardless of our cultural or intellectual background, theoretical commitments, or indeed an ability to think at all. Relatedly, although admittedly for some less convincingly, I believe that we can also say that states of positive valence have intrinsic value. In other words, pleasurable states such as happiness, bliss, and joy are good in and of themselves. They have an inherent radiant value—an "ought-to-be-ness."

As we will explore, it is in our direct acquaintance with these intrinsic qualities that we discover unequivocally that morality is not, as many have argued, simply a matter of evolutionary psychology, opinion, or tradition, but is in fact a discernible aspect of the universe, as real and consequential as the force of gravity. I will call this type of view "Conscious Value Realism" and it finds echoes in the work of, among others, the philosophers David Pearce[6], Magnus Vinding[7], Thomas Nagel[8], Sam Harris[9], and Peter Singer[10]. But it is perhaps most clearly articulated in the work of Sharon Hewitt Rawlette, whose concept of *normative qualia* provides one of the clearest accounts of the core insight this book explores: that certain conscious states have an inherent ethical significance and provide the foundation of a truly rational moral realism.[11] Together with these and other emerging voices, I will suggest that Conscious Value Realism has the potential to bring profound new clarity to the most urgent ethical realities of the world.

This is a journey that will take us beyond the realm of theory, from the hidden hells of factory farms to unimaginable peaks of conscious flourishing; from your immediate experience in this very moment to conscious states occurring trillions of years in the future; from the world-changing potential of superintelligent AI to the possibility of truly conscious artificial minds; from the sentient inner lives of octopuses and nonhuman animals to alien minds in the depths of space. In this book we will explore an emerging moral realism with the power to make sense of the unique and extraordinary historical moment in which we find ourselves, the vast moral realities that loom all around us, and choices of truly cosmic significance.

2. MORAL OBJECTIVITY

It might sound strange—even contradictory—to suggest that the deeply private and subjective texture of conscious experience could be the foundation of an objective moral realism. After all, the words "subjective" and "objective" appear to point in opposite directions, picking out two very different parts of the universe. And yet this does not mean that there cannot be objective facts about subjective states.

For example, it is objectively true that I am currently experiencing a mild discomfort in my left knee as a result of an old skateboarding injury. Here, I take the word "objective" to mean a truth that is independent of the beliefs and judgments of observers. My experience of this pain is a truth about the universe. The fact that this experience is, for the most part, private to me does not make it any less real. The crux of my position, however, rests on the further claim: that it is also an objective fact that it would be better if I did not suffer this pain. This is an objective fact that is not merely true for me; it is, I believe, true in the nature of the experience *itself*.

In other words, if we attempt to objectively describe all of the characteristics of an experience of suffering, no account would be complete that does not make reference to its inherent "dissatisfactoriness," its "ought-not-to-be-ness." The felt reality of these experiences provides an objective motivation—to act on their value: that, all things being equal, we have a rational reason to minimize states of suffering and to promote states of happiness and flourishing.

At this point, readers familiar with the study of morality and ethics (I use these words interchangeably) may well be wondering what David Hume might have to say about all this talk of moral realism. Hume, for those who may not be familiar, was a well-liked and respected 18th-century philosopher who was famously skeptical of moral realism. His rejection of the idea has been so persuasive that his widely known dictum "You cannot get an ought from an is" is often regarded to be the last word on any discussion of moral realism. Hume's observation was that when we examine the facts of the world, we never discover anything resembling good or bad. The world just *is*—an indifferent collection of facts. We can search high and low, but no matter how hard we look, nowhere, Hume argued, does the world ever prescribe what we ought to do. Thus, according to Hume, when people make claims about morality, they are, in fact, merely expressing their opinions, preferences, and feelings about certain behaviors and outcomes.[1]

As much as I respect Hume, I think he was, on this matter, deeply mistaken. What Hume overlooked was that, contained within the universe of facts, there also exist facts about conscious states, and that certain among these states have an inherent moral significance. Suffering, again, is the prime example. The ought-not-to-be-ness of suffering is not something that we infer about the experience; it is on the surface of the phenomenology of the state itself. It is part of a purely factual explanation of suffering as a real conscious state occurring in the universe. Recognizing that this ought-not-to-be-ness is inherent to suffering forces us to recognize that from the fact of suffering, an "ought" does, and indeed must, follow.

Let's pause and spell this out in simpler terms, because Hume's famous rule—"you can't get an ought from an is"—can feel like an unbeatable trump card. Hume is right: if you begin with value-free facts, any claimed moral duty appears conjured out of thin air. The response, however, is that suffering and joy were never value-free to begin with. The felt badness of burning your hand on a stove is baked into the experience itself, just as heat is baked into the metal. Anyone who feels that pain already knows, without argument, that it should stop. That little "should" lives *inside* the state; you don't add it on afterwards, and you don't need a philosopher to certify it. In other words, the "ought" is already nested within this particular "is," so Hume's gap never opens in the first place.

This, I believe, is the point Hume missed, that in the context of these conscious states, we can, and indeed have no choice but to derive "ought" from "is." Unfortunately, the widespread belief that Hume decisively closed the door to moral realism has prevented the development of a rational and evidence-based discipline of ethics that could build upon the judgment-independent value of conscious states. I think we can view this as an unfortunate setback for our civilization, and one that we must now urgently address.

I will be defending moral realism on the basis of value realism. For some thinkers, this type of view smacks of a metaphysical naivety about value—which is to say that it's a view that may appear to rely on vague or "spooky" notions about how value exists in the world. It is widely believed that value exists only as a psychological impression and not something real or discoverable about the "real" world. This is partly fair, because the value in question arises exclusively in consciousness, and yet who said that conscious states are not real?

Part of the confusion stems from the strange place that consciousness has come to occupy in our academic and scientific frameworks. In the hierarchy of existence, we have come to view the physical world as preeminent and any subjective phenomena as the mere vaporous denizens of an illusory domain—shadows on the wall, less than real. But consciousness is undeniably real, and the fact that it remains unexplained at the level of physics and biology does not undermine its reality. In fact, consciousness can be rightly viewed as having a higher form of existence beyond anything purely physical. This is not a deep claim about the fundamentality of consciousness—though that is, I think, a live possibility. Instead, it is the recognition that mental states have the potential of truly mattering in a way that nothing unconscious ever can. As the philosopher-historian Yuval Noah Harari observes,

> "If you really want to explore reality, the best place to start is with suffering—because it is the most real thing in the world."[2]

It is in this sense that I am claiming moral realism is true: that there is a real, discoverable space of ethics that is entailed by the existence of, and the potential for, these states. That just as the conceptual existence of the numbers one and zero entail a larger rational space of mathematics, the mental states of pleasure and suffering, broadly construed, entail a larger, rational space of morality.

A relevant observation is that even in a state of non-existence, certain facts still hold true. Mathematical truths are a good example. We don't appear to need a Big Bang or a universe for it to be true that one plus one equals two, or that there is a specific set of prime numbers. There is an in-depth philosophical discussion around the correct form of reality that we should assign to such mathematical truths, and yet to recognize that these truths exist in a real and consequential way is not an appeal to a metaphysical naivety about mathematics. Conscious Value Realism holds that similar truths exist about value: that even in a state of non-existence, it is nevertheless true that conscious suffering, were it to come into existence, is bad. Similarly, conscious wellbeing, broadly defined, is good. Just as mathematical truths are universally valid and exist in a state of abstract objectivity, there are truths about value that hold in any conceivable universe or state of existence.

An important feature of this view is that value is a phenomenon which arises in conscious experience. For those who may agree that value does exist, the assertion that it exists exclusively in consciousness may still be contentious. There are various ethical systems that project intrinsic value beyond the realm of consciousness into such things as the natural world, the diversity of species, or the living biosphere itself. These phenomena, it is argued by some, hold inherent value, independent of any conscious observer. Such thinkers might argue that it is implausible to claim that a majestic ancient forest holds no intrinsic value beyond that which is realized in its conscious onlookers and inhabitants.

But is it really plausible, or indeed intelligible, to claim that value could exist in isolation of any possible conscious state—either now or in the future? There's no question that the forest has immense value, and yet I believe we must recognize that this value exists only in its potential realization in consciousness, and that the notion of value without a valuer, without consciousness, simply doesn't make sense. Without it, there is no joy, no suffering, no preference, no aversion, no meaning, no good, no bad, no perspective, no subjectivity—no anything that matters. Strip away consciousness, and the universe is just inert particles interacting blindly, with no stake in any outcome.

Despite this, we humans wrestle with a strong intuition that value does reach outward beyond consciousness. The basis of this intuition has a deep-rooted evolutionary origin. We evolved to project value into the external world—into the objects of our desires and into the environmental conditions of our survival. In reality, however, I believe we can view value as similar by analogy to color: that just as it took thousands of years for humans to finally understand that colors exist exclusively in our perceptions, as opposed to dancing on the surfaces of objects, we will eventually recognize that value is a phenomenon arising in experience and that it cannot intelligibly exist apart from it.

This is an important point, and we'll return to it later. But we'll now turn to another important question: how can suffering be intrinsically disvaluable when it appears to serve so many vital functions? Similarly, how can pleasure be intrinsically valuable when it so often leads us astray?

3. INTRINSIC AND INSTRUMENTAL VALUE

Over millions of years and countless generations, natural selection has recruited suffering as a signal—to warn us of danger, to steer us from injury, and to inform us more generally when things are not going well. It is a mistake, however, to confuse this practical or "instrumental" value with intrinsic value. While, as I claim, suffering is intrinsically disvaluable—bad in itself—it can simultaneously have great instrumental value: in causing us to avoid greater potential suffering down the line.

People occasionally say things like, "No pain, no gain" or "If you never fall, you never learn." Of course, there is some truth here, where a certain amount of suffering proves instrumental in the attainment of some desirable outcome. Sometimes an acute confrontation with what we really don't want can focus our mind, perhaps for the first time, towards what we really *do* want, causing us to make vital steps in that direction. All too often, however, we confuse this instrumental value with intrinsic value. After all, if the same positive outcomes could be achieved without any of the suffering, we would have no reason to go through it. Furthermore, to do so, I believe, would be a kind of moral error.

Similarly, we are all familiar with times when the allure of pleasurable states results in seemingly bad outcomes. An example might be when we are tempted by the pleasurable taste of an otherwise unhealthy food. Eating delicious but unhealthy food can be considered "bad" in the instrumental sense—because poor health will tend to cause suffering and curtail wellbeing in the long run. But the pleasant experience of eating the food is good in the full intrinsic sense—yet this is a measure of good that also involves a compromise with the quality of conscious states down the line. The claim, then, is that while it may be unwise to indulge in unhealthy food, the pleasure of eating it is nevertheless inherently valuable.

In the interest of testing the limits of this view, it might be helpful to consider an extreme example that will challenge our intuitions. *Sadism* is a trait in which people derive pleasure from causing suffering to others. In Conscious Value Realism, the pleasure felt by the sadist as they torture their victim is intrinsically good. To most of us, this claim feels very wrong. How could pleasure have intrinsic value under such circumstances? I believe that our intuition against this is yet another example of the difficulty we often have in distinguishing between instrumental and intrinsic value.

The repugnance we feel towards the sadist is not appropriately directed to the pleasure itself. Instead, our concern is more correctly directed to the distinctly zero-sum manner by which this pleasure has been brought about. In our repugnance, we are also responding, I think, to the kind of person that would behave this way, the associated danger of cooperating with such an individual, and what else for which they might be culpable.

While intrinsically valuable, the pleasure of the sadist is extremely disvaluable in an instrumental sense because it is rewarding and thereby promoting the creation of suffering. Indeed, sadism is doubly bad because it fails to recognize the impartial nature of intrinsic value: that pleasure and suffering are intrinsically valuable and disvaluable regardless of where, when, or in what they are instantiated. This "impartiality" is a key feature of Conscious Value Realism that we will explore in more detail later.

Another point of potential confusion is the distinction between sensations and suffering. Consider, for example, the familiar burn of vigorous exercise. The very same throbbing or burning sensations in our muscles that might feel euphoric in the gym could, in another context, feel intolerable. What is distinct between these experiences is not the physical stimulus but the valence—the feeling tone—that accompanies it. Suffering is not the sensation itself, but the quality of "ought-not-to-be-ness" or *dissatisfactoriness* that may or may not accompany it.

As Rawlette has pointed out, strong evidence for this comes from reports of patients who are given opioid analgesic medication for the treatment of chronic pain.[1] Many describe that, while they still feel the physical pain, they simply no longer care about it. The bodily signals remain. Yet what disappears is the badness. This suggests that opioids actually target the felt quality of disvalue itself, as opposed to merely muting the sensory input. That opioids act directly on what Rawlette calls "phenomenal badness" may well go some way to explaining why they are among the most addictive known substances.

Another example we might consider is the case of sadomasochists, individuals who report an enjoyment of the experience of pain. At first glance, the existence of such individuals might seem to challenge the claim that suffering is intrinsically disvaluable. And yet, a closer look shows that to draw this conclusion would be to once again mistake sensations for suffering. If an individual reports pleasure and arousal while engaging in painful acts, it merely reveals that for these individuals, these sensations elicit positive rather than negative valence.

Suffering, as we've defined it, is not merely a sensation, but a specific negative quality of conscious experience. Its disvalue does not depend on judgment, interpretation, or context—it is directly given in the experience itself. For the sadomasochist, at least under certain conditions, pain sensations do not produce this negative valence, and so no intrinsic disvalue is present.

Such tendencies may be the result of prior associations, psychological conditioning, or neurological differences that cause pain sensations to acquire a positive valence. But this does not undermine the claim that suffering is intrinsically disvaluable—because in these cases, suffering simply is not occurring.

In fact, the existence of sadomasochists underscores the judgment-independent nature of conscious value. They do not choose to have this trait, and the behavior it promotes often carries obvious risks: a heightened chance of injury, or difficulties in empathizing with the suffering of others. All things considered, if one had the choice to become a sadomasochist, doing so would be irrational for practical reasons. Pleasure remains intrinsically good, and suffering intrinsically bad—even when they arise in unexpected forms or contexts.

Another possibility to consider is our seeming ability to influence how we respond to sensations so that, for example, we can avoid experiences of suffering. Speaking personally, I've needed to have several wisdom teeth removed in recent years. While undergoing these procedures, I've attempted, with varying success, to prevent certain sensations producing suffering in my experience. In situations like these, many of us will instinctively attempt something similar; to control our breathing, to contextualize the sensations of pain by gently reminding ourselves that the procedure is all for the good, and will be over soon enough.

This is, of course, a valuable skill to have, and one that is worth cultivating. It is important to understand, however, that this capacity does not suggest that suffering is merely an expression of personal preferences, or that suffering itself is not intrinsically disvaluable. The contextual significance of sensations, which can but do not always produce suffering, shows instead that sensations and suffering are not identical—that phenomenal badness or goodness are distinct qualities that are either a part of an experience or not.

It's important to acknowledge a source of potential confusion that arises from the wide diversity of circumstances under which we experience suffering and pleasure. The many conditions in which we experience these states can give us the impression that they are so varied and different from each other that they are in fact *categorically* different—that while we might use the word "suffering" to describe a great variety of states, in reality some of these states have an altogether different experiential nature. For example, the suffering of banging one's knee on the edge of a door feels to be of an altogether different character to the suffering which might accompany a bereavement. This is to be expected, given the vastly different contexts and physical mechanisms involved in these experiences. The intense jolt of physical pain and subsequent throbbing of banging one's knee seemingly has almost nothing in common with the malaise of sadness—the aching sensation in the chest and solar plexus that is attendant to the loss of a loved one.

While the phenomenology (or feeling) of these states differs vastly, for our purposes, it is what they have in common which must not be overlooked: the experience of phenomenal badness. *This* is what we are calling "suffering," and while we can experience it under a spectrum of different circumstances, I would argue that under careful introspection, we can recognize that it is a common and distinct quality of experience—so fundamental that we often overlook it.

As usual, a mirror version of this view extends to pleasurable states or "phenomenal goodness." The positive quality of experience that is attendant to enjoying one's favorite food, for example, feels very different to the experience of falling in love. Nevertheless, there is a distinct and common quality of goodness, a positive valence, that runs through both of these experiences.

A related discussion surrounds the difficulty of comparing different forms of pleasure and suffering, and the lack of any clear quantification. How many pinpricks should I be willing to endure to avoid a migraine? Precise answers to such questions may seem impossible. In reality, however, I suspect that at least some such answers do exist, but that they are simply very difficult to acquire. Indeed, it is conceivable that a future science of consciousness could include a mathematical formalization of valence—a framework that in certain contexts could offer real answers about the relative moral weight of different experiences.

Such a theory of valence would be a transformative breakthrough. Imagine, for example, that through some technology based on this breakthrough, we could actually *see* a person's real-time valence, translated perhaps into color, appearing around the individual like an aura. In this visualization tranquil, flowing blues may convey states of deep peace or euphoria, while stark, jagged reds convey intense fear and suffering. Such a display may leave out much, if not most, of what we truly care about, and yet to imagine a world in which the valenced states of others could, if desired, be indirectly perceived, is to imagine a world transformed, in which certain moral realities are brought into undeniable focus.

We will return to this point later, and yet as we shall see, for the most urgent moral realities of our world, such precision is not required. It is enough, in many cases, to understand that certain actions produce great amounts of needless and avoidable suffering, and that there are actions we can take to prevent it.

Before we move on, we should also note that there have been philosophers, both historical and contemporary, who argue specifically against the claim that suffering is intrinsically bad and that pleasurable states are intrinsically good. It has been argued, for example, that rather than being truly disvaluable, states of suffering actually exist more like a symbol or representation of disvalue, and that while these states may alert us to something that is potentially dangerous, the state *itself* is not bad. In other words, suffering isn't disvaluable—it merely *represents* disvalue. Pleasurable states are not good *in themselves*; they merely symbolize goodness for us.

Part of what makes the view tempting is that our conscious states often arise in familiar and meaningful contexts: we feel pain when injured, or joy when reuniting with a loved one. These associations, shaped by evolution, can make it appear as though the value lies in the external event, rather than in the experience itself. But this is an illusion. Upon deeper reflection, the view collapses—not merely because it fails to align with the reality of experience, but because it is conceptually confused.

Firstly, it is simply not how suffering presents itself in experience. Suffering does not come to us as a representation of badness—it *is* the badness. When we are overwhelmed by grief, or writhing in agony, or tormented by despair, the sense is not that we are observing a symbolic warning of something else that is bad. The *experience* is what is bad. To be in pain is not to observe a signal, like a dashboard light—it is to be immersed in something intolerable. The very phenomenology or feeling of the experience defies this distancing. In fact, if we were truly able to perceive our suffering merely as a symbolic alert—detached, abstract, semantically neutral—it would cease to be suffering at all. It would lose the quality of phenomenal badness.

The same is true of positive experiences. States of joy, peace, or euphoria do not appear to us as mere indications that something good has occurred. They *are* the good. When we are immersed in delight or wonder, we do not experience these states as signs pointing elsewhere, but as the very realization of goodness itself. It is not the symbolic function of these states which makes them desirable—it is their felt reality. Like suffering, pleasure is not a messenger of value—it is the *bearer* of it.

It is also worth noting that to suggest that suffering merely represents badness commits a kind of conceptual inversion. It presumes that the badness of injury or loss is somehow more basic than the badness of suffering. But what is it, precisely, that makes injury or loss bad, if not the suffering it causes or entails? Strip away the conscious disvalue—remove any distress, grief, or diminution of conscious wellbeing—and the notion that a physical injury or personal loss is intrinsically bad becomes empty. We care about these states only insofar as they determine the quality of conscious experience either now or in the future. Suffering does not represent badness—it is the reason we believe there is badness in the world in the first place. It is the foundation from which we derive all other notions of harm.

The mistaken view—that suffering is a kind of representation or symbol rather than an intrinsic feature of experience—illustrates the dangers of overintellectualizing what is directly given in conscious experience. It is a view that could only be held by someone at a great philosophical distance from the lived reality of suffering or joy—someone who has, in a sense, turned away from the very experience itself in favor of conceptual abstraction. This is philosophy at its most estranged. We do not need a theory to know that joy is good and agony is bad. Their value is self-revealing. Suffering discloses itself as something that ought not to be; pleasure as something to be promoted. To deny this is not just to make an error in theory—it is to turn away from the very light by which moral awareness becomes possible.

Other claims that pleasure and suffering are not intrinsically good and bad, often fall within one of two categories. They either fail to make the distinction that we have already drawn between sensations and suffering, or they instead argue that suffering, such as that occurring as a result of a bereavement, is a "natural" and necessary psychological process along the path to healing. This argument makes the now familiar error of confusing instrumental and intrinsic value. It would certainly appear to be the case that, as human beings, suffering is a deeply integrated element of our psychological makeup—a core feature of the mind's functioning. If we felt no negative emotion at the loss of a loved one, we might be right to worry that something deeply important is missing.

And yet this does not in fact challenge the view that suffering is intrinsically disvaluable. For beings such as we, states of suffering may well lie in the path between ourselves and a future that involves more positive conscious states, but the fact that suffering may in some cases be unavoidable is ultimately circumstantial—an inherited feature of our evolutionary psychology, and not, as it might seem, a challenge to the intrinsic disvalue of suffering. Natural selection, after all, did not shape our minds under the guidance of an ethics committee. For billions of years life has been a process that responds to only one signal: reproduction.

Recognizing this fact brushes up against an uncomfortable truth that, sooner or later, we must be willing to confront. It is that, to the best of our knowledge, natural selection recruited consciousness purely in service of successful reproduction. Darwinian evolution is not, and has never been, in the business of creating happy minds. As Homo sapiens, to be truly happy, we must fully placate the whims of our evolutionary psychology. And even then, our evolved minds impose an upper limit to how happy any of us can be as well as how much suffering it is possible to avoid.

As Darwinian creations, we find ourselves on what the philosopher David Pearce has called "the hedonic treadmill"—forever seeking temporary respite from craving, constantly adapting to our levels of happiness and suffering, perpetually striving but never truly finding lasting contentment.[2] On reflection, we can recognize that much of our evolved psychology is distinctly out of alignment with what a system that is optimized for conscious wellbeing would look like.

There is a challenging conversation to be had about Darwinian natural selection—as a brute mechanism of replication that recruits consciousness into servitude to that end. Ultimately, our minds are the product of an apparently blind evolutionary process driven solely by survival, and one day it might be for us to decide if the mind it has bestowed upon us is the one that we really want. We'll return to this thorny issue later, as well as the promise, perils, and pitfalls it presents.

4. HEDONIC VALUE AND IMPARTIALITY

> "For reasons we do not understand, the pain-pleasure axis discloses the world's inbuilt metric of (dis)value."[1]
>
> -David Pearce
> Philosopher

Given our focus so far, the question naturally arises: Are positive and negatively valenced experiences (also known as "hedonic" experiences) exhaustive of all intrinsic value and moral significance? Might there be other states in the universe that are objectively good and bad in themselves? For some people, the claim that hedonic states are what ground moral significance feels deeply wrong. What, they might ask, about the vast range of experiences which make up the rich tapestry of life? What about when the pleasures and pains of our experiences are layered and intricately interwoven—the beautiful sadness of nostalgia, or the aching melancholy at the heart of love? Much of what is interesting and apparently valuable about life's experiences can appear, at least initially, to have little to do with pleasure and suffering. To reduce the value of our inner lives to mere valence would appear to overlook much, if not most, of what we truly care about.

The first point to clarify is that, while we are claiming that hedonic states have intrinsic value and thus moral significance, we are not claiming that they are the only phenomena in the universe that possess this quality. Perhaps there are other forms of intrinsic value. And yet, the recognition that hedonic states are an example of such value is all that is required to establish a foundation for moral realism. The sighting of a single magpie is enough to prove that birds exist, even if we remain unsure whether swans, penguins, or parrots do as well. Likewise, identifying the intrinsic value of hedonic states is sufficient to show that moral value is a real feature of the universe, even if other such values remain undiscovered.

While suffering and pleasure are by far the most conspicuous ways through which intrinsic value and moral significance arise in the universe, I believe we should remain open-minded to the possibility that other forms of intrinsic value exist to be identified or discovered. After all, recognizing the moral weight of valenced experience was itself such a discovery, and justifies our remaining open to other potential forms of intrinsic value. If such states exist and we fail to recognize them, we could simultaneously overlook vast realms of potential value and in doing so become the unwitting authors of moral catastrophes. Indeed, as we will explore later, our collective failure to recognize the intrinsic disvalue of suffering has produced and prolonged various ongoing moral emergencies that we must now urgently address.

For the moment then, let us entertain the possibility that, beyond positive and negative hedonic states, other, as-yet-unrecognized forms of intrinsic value *do* exist. But what might such states be? Might they already be familiar to us? For a species of philosopher known as the "virtue ethicist," there are several candidates ready and waiting in the wings. The virtue ethicist might argue, for example, that aesthetic beauty has intrinsic value: that all on its own, beauty just makes the world better. Certainly, there can be no doubt that beauty holds great value. Yet I would argue, however, that this value is plausibly exhausted by its instrumental ability to elicit positively valenced experiences in various ways. Beauty can be evocative of creativity, new ideas, and deeper thought and reflection, but the end value is always plausibly reducible to the positively valenced states that beauty can evoke. To be clear, this may not be the full story about the ultimate value of beauty, but I would claim that we don't have any foundational basis by which to infer intrinsic value to beauty in the way that we do have in our direct acquaintance with pleasure (or suffering, in the negative case).

Among the most persistent candidates for non-hedonic intrinsic value is the idea of *meaning*: not just pleasurable or painful experience, but the sense that one's life, or some part of it, participates in a larger structure of significance. This intuition has deep roots. Philosophers including Susan Wolf[2] and Thomas Nagel[3] have argued that meaningfulness is a kind of value in its own right: not reducible to pleasure, not synonymous with virtue, but something which matters independently.

I believe, however, that on closer examination, the appearance of meaning as an intrinsic value dissolves. Just as with other virtues, the moral weight of meaning plausibly derives from its capacity to promote and protect wellbeing. What we call "meaning" is less a distinct form of value than a cognitive and emotional lens that can bring coherence, intelligibility, and narrative fit to what might otherwise be confusing, painful, or directionless.

The presence of meaning transforms the tone of experience. A loss made meaningful is less corrosive. A sacrifice understood as worthwhile becomes bearable. To feel that something "matters" is, at root, to feel a shift in the valence of experience. Meaning permits the inclusion of even difficult or painful experiences within a larger sense of goodness: that it was worth it, that it mattered, that it belongs.

This dynamic is not limited to the past. Meaning can prevent suffering from arising in the first place. As in my earlier example of having wisdom tooth surgery, when pain is framed as purposeful, necessary or worthwhile, it can remain as mere pain rather than transforming into suffering. The mind's interpretation, its sense of significance, determines whether a sensation corrodes wellbeing or remains endurable.

Meaning also amplifies positive states, imbuing joy or insight with a felt sense of order, resonance, and connection. But meaning is not inherently benign. A destructive ideology can give life a sense of overwhelming purpose while licensing cruelty or self-destruction. A false belief in a conspiracy theory can render random misfortune "intelligible" while deepening fear, alienation, and detachment from reality. The possibility of negative meaning further challenges the view that it is intrinsically valuable.

In all these cases, the moral significance of meaning appears to flow from valence: enriching joy, contextualizing pain, preventing suffering, or sometimes amplifying it. Meaning enhances wellbeing not by introducing a new category of intrinsic value, but by shaping how experience feels from within. This understanding of meaning does not diminish its importance. Meaning may be among the most powerful psychological tools for sustaining and cultivating wellbeing, yet its value is best understood as instrumental rather than intrinsic: a vital cognitive structure, but not an end in itself.

Numerous virtues are often hailed to be intrinsically valuable. They include, among others, courage, fairness, loyalty, and wisdom. While we often speak about these virtues as though they are intrinsically valuable, on closer inspection, their value appears to be, in every case, subsumed by their instrumental potential—as strategies for engineering positive conscious states and avoiding negative ones. Fairness, for example, is instrumentally valuable in numerous situations, but we should not fail to recognize that a blind commitment to fairness could entail the perverse goal in which everyone must suffer the same. Fairness, therefore, cannot be a good in itself. Similarly, the virtue of loyalty can have great instrumental value; however, it is not itself valuable if this loyalty is to a dangerous psychopath. Finally, we might ask what good wisdom would have if it consisted of no path away from suffering or toward some lasting wellbeing. It would seem that all apparent virtues can be corrupted when they are divorced from their capacity to elicit positive or negatively valenced experiences.

Such an understanding does not undermine the great instrumental value of "virtuous" strategies and behaviors. It seems likely that, all things being equal, it will always be preferable to cultivate rather than stifle creativity. Cooperation will always be preferable to conflict, integrity over dishonesty, freedom over enslavement, knowledge over ignorance, and reason over chaos. And yet when we reflect upon them, it is apparent that, just like fairness and loyalty, each can be deployed at the wrong time or in the wrong context in a way that produces negative results. While their value may not be intrinsic to them, this does not mean that these revered states are anything less than indispensable—enduring strategies that promote flourishing across a diversity of minds.

Ultimately, "virtues" like truth are often distinctive because of the high instrumental potential they represent. A possible evolutionary explanation is that our inclination to view certain virtues as intrinsically valuable is a kind of cognitive shortcut that we evolved which removes a need to recall all the ways a particular action can produce positive outcomes. An example is telling the truth. Much of what we care about would fall apart if we could not generally rely on others to tell the truth, and so we tend to just view telling the truth as inherently good. And yet when pushed, we can readily imagine situations when telling the truth will produce bad outcomes. The classic example is the Nazi officer searching for information about Jews hiding in the local area. Under such circumstances, a convincing lie could mean the difference between life and death. On reflection, it seems likely that the ultimate worth of virtues like truthfulness, loyalty, fairness, and wisdom derives entirely from their capacity to generate positively valenced experiences and avoid negative ones.

All this being said, many of us come from a different ethical starting point. Perhaps you feel, for instance, that some actions are simply right or wrong regardless of what the outcome might be, or that certain virtues carry a kind of intrinsic dignity. If you feel this way, then you're in the company of one of the most influential moral thinkers of all time: the 18th-century philosopher Immanuel Kant. For Kant, moral law was rooted not in outcomes but in rational duty, and some actions—such as lying—were categorically wrong, no matter the consequences.[4] This is known as a "deontological" moral view—*deon* coming from the Greek word for "duty." Kant's framework led him to argue that it would be wrong to lie under *any* circumstance.

Under this view, if the Nazis are at your door, you must tell them if asked about the Jews hiding nearby. For many, this reveals both the power and the limits of Kantian rigor: it captures the intuition that some principles are morally inviolable, but also shows that strict adherence to them can lead to morally troubling outcomes. If you agree that lying would be permissible under these circumstances, then you agree that the virtue of telling the truth is subject to certain exceptions. This is not to say that virtues like telling the truth should be discarded, but that telling the truth is not an end in itself. The true end, Conscious Value Realism suggests, is how our actions impact the space of consciousness.

In reality, the view we are defending doesn't simply explain away these intuitions; it grounds them. If certain principles or character traits feel morally indispensable, it is likely because they reliably shape the quality of conscious life—for ourselves and others. We are not mistaken in caring deeply about fairness, honesty, or responsibility. And yet, Conscious Value Realism suggests that what gives these qualities their moral weight isn't just a matter of raw intuition, but something *real*—how they tend to influence the texture of consciousness: to preserve those qualities of experience that are unambiguously good, and to steer us away from those that are truly disvaluable. The importance of these virtues is not diminished by viewing them through the lens of Conscious Value Realism. If anything, their enduring role in our moral life is clarified by it.

Earlier we defended the view that value is a phenomenon that arises exclusively in consciousness, and that to invoke the concept of value in a way that is separate from any conscious experience is to misunderstand what value fundamentally is—namely, a quality that discloses itself through the felt character of experience. Without consciousness, the very concept of value becomes empty and incoherent. Nevertheless, we might still plausibly ask if there could be other conscious states that are intrinsically valuable but which are not hedonic in nature.

To consider this possibility, it is worth reflecting that the human mind has access to what may ultimately be only a tiny region of the total state space of possible conscious experience. Vast realms of consciousness undoubtedly await to be discovered, perhaps by our descendants, whose minds may far exceed our current capacities and sensitivities: music we lack the ears to hear, beauty we cannot yet see. The philosopher Toby Ord expresses this sentiment in his book *The Precipice*, where he writes:

> "Consider, for example, how little we know of how ultraviolet light looks to a finch; of how echolocation feels to a bat, or a dolphin; of the way that a red fox, or a homing pigeon, experiences the Earth's magnetic field. Such uncharted experiences exist in minds much less sophisticated than our own. What experiences, possibly of immense value, could be accessible, then, to minds much greater? Mice know very little of music, art, or humor. Toward what experiences are we as mice? What beauties are we blind to?" [5]

If Conscious Value Realism is true, an advancing civilization may rationally converge on the project of systematically exploring the vast state space of conscious experience and the value which can be realized therein. As the philosopher-historian Yuval Noah Harari puts it:

> "There is an entire ocean of consciousness, of mental states out there that we know nothing about. Some of them have existed before in other animals, but many of them have never existed in any animal. But they are still part of the possible spectrum of mental states. It is likely that in the coming decades and centuries, we will start exploring this ocean of consciousness, like the European explorers of the 14th and 15th centuries started to explore the oceans of planet Earth."[6]

Such an odyssey may reveal profound and as-yet-unimaginable forms of conscious understanding and ways of perceiving—alien mind spaces and radically new species of qualia (or feeling), as distinct from everything we know as the faculty of sight is from hearing. Within this near-infinite ocean, might there exist other forms of intrinsic value, that have nothing to do with hedonic valence? Is it really true that all possible mental states can be intrinsically valuable only insofar that they hum with the presence of felt goodness or felt badness?

This claim is undeniably bold, and given the stakes, we should be very cautious about accepting it. "Locking in" a purely hedonic view of value could blind us to entire realms of moral significance. That said, if we accept that value arises solely in consciousness, and that we come to know this value through direct experience, then it's very difficult to see how any conscious state could be intrinsically valuable without feeling that way from the inside.

To experience something as good or bad—to *feel* that it matters—is to be in a positively or negatively valenced state. It would seem that if a conscious state doesn't reveal its value through how it is subjectively felt, then its value cannot be intrinsic to the experience itself—it's something inferred or projected onto it from the outside. This makes the case for non-hedonic intrinsic value more difficult: if a conscious state does not present itself as good or bad in the experience, then in what meaningful sense can we say it is good or bad in itself?

Again, given the moral risk of potentially overlooking entire dimensions of value, the jury remains out on the existence of such values, and perhaps should remain so indefinitely. Still, hedonic valence remains our most concrete and directly knowable basis for intrinsic value. And yet, many people express a certain allergy to a hedonistic view of value that strikes me as misplaced—often stemming from a misunderstanding of what hedonism actually entails. In popular usage, "hedonism" has taken on connotations that bear little resemblance to its true philosophical meaning.

For example, it does not, as often depicted, endorse an approach to life as though it were one long party, without consideration of others or future consequences. This popular depiction mischaracterizes hedonism in three important ways. The first is that hedonism is all about immediate gratification with no thought for the future. In fact, hedonism, in its pure form, does not assume that states in the present are any more important than those that will arise in the future. In reality, recognizing that these states embody intrinsic value demands deep reflection about how the future might unfold and will often require trade-offs between present and future conscious states.

Secondly, picturing a hedonistic view of value as advocating a life of Dionysian thrill-seeking and base pleasures is a caricature of what are, in reality, the vast depths of experience encompassed by the broad categories of positive and negative states. In addition to base pleasures, like those we associate with sex and eating, there exists a vast universe of potential valenced experience—subtle, textured, interwoven, expansive, and even spiritual—each with intrinsic value disclosed through its felt quality. Deep and varied forms of satisfaction, comprehension, and meaning; exquisite modes of qualia both known and as yet undiscovered. And the reverse, of course, is true of negative states: all possible forms of harm and breaches of wellbeing—experiences that, by their very nature, we are right to avoid.

The third, but most significant misconception about a hedonistic view of value is that it is a selfish view, concerned only with one's own experiences. Philosophical hedonism, however, asserts that *all* valenced states have intrinsic value, regardless of the subject in which they arise. In its pure form, the view makes no distinction about where such states are occurring, be they in ourselves or in any other conscious being. Indeed, to recognize the value of a conscious state as being intrinsic to it, hedonic or otherwise, is to view this value as independent of all other factors, including its location in space and time, whether it arises in ourselves or in another species, in other races, genders, and so on.

This impartiality emerges as a core insight of Conscious Value Realism. Someone might plausibly ask: Why should we care about any experiences beyond those that are, or will occur, within the theatre of our own minds? Why not seek to promote only our own happiness, even when it is at the expense of others? As Rawlette adeptly points out, to know that something is objectively good is to have a reason, and indeed an 'ought,' to promote it, whether we are directly acquainted with this value or not.[7] In a strict sense, we therefore have the same rational basis to promote happiness and to avoid suffering in others as we do in ourselves.

Of course, our own conscious states loom far larger to us than the merely abstract idea of experiences arising in others. This makes it very difficult for us to give the appropriate moral weight to experiences that we are not directly acquainted with and never will be. But this, it would seem, is simply a result of the kind of being that we are. If we were instead what philosophers have called "ideal observers" (that is, if we were fully rational and in full possession of all relevant facts) then we would know directly that all like experiences carry equivalent value and thus moral weight. To care only about one's own experiences is therefore to be in the grip of an epistemic illusion—to act irrationally, from a position born out of ignorance. This important truth hints at a broader discussion about identity and the nature of the self that we will explore later in this book.

In seeking to make rational ethical judgments, we are at a disadvantage as a result of our epistemic isolation from other minds. Again, we only truly feel the full value and moral weight of our own experiences. For the most part, the conscious states of others exist to us in a merely conceptual and abstract way. Practically speaking, this isolation is beneficial to us, since direct knowledge of the uncountable experiences occurring in reality would, of course, be an overwhelming psychological burden. Nonetheless, with moral rationality as our goal, our task is to try to think more like ideal observers: recognizing the intrinsic value and moral significance of conscious states, wherever and whenever they may be arising in the universe.

It is true of course that the practicalities of human life make absolute impartiality unrealistic. But absolute impartiality is not what is called for. Each of us has what we might call "special duties" toward particular conscious beings. These special duties are defined by the greater reach of our influence into specific theaters of consciousness, and they account for many of our familiar relationships. For example, even though the like experiences of any given child can be seen as having equivalent ethical weight, under most circumstances, a mother is morally justified in prioritizing the needs of her own child. This is because only she, and perhaps a few others, will ensure the wellbeing of this specific child.

This framework of special duties extends to many aspects of human relationships: friendships, partnerships, caregiving, and community ties, all of which are shaped by our ability to impact the lives of particular individuals in ways that others cannot. Most of us have numerous such special duties, and they are all defined by the differing reach of our influence into the conscious experiences of specific minds. As we will continue to explore, our moral responsibility is broadly defined by the reach of our influence: by how our actions affect the conscious experiences of others, whether directly or indirectly, now or in the future. In other words, the reach of our influence and the scope of our moral responsibility are one and the same.

While ethical impartiality challenges the instinctive prioritization of oneself over others, it does not demand a complete abdication of self-interest. Notably, we have an important special duty to ourselves. After all, we too are moral patients capable of experiencing morally significant states, and few others will affect these states more than we do through our own choices and actions. Thus, the moral responsibility to ourselves is both inevitable and legitimate, as it reflects the same ethical principles that guide our concern for others.

Recognizing the moral significance of all conscious beings does not require abandoning our responsibilities to those whose lives are most closely entangled with our own. In a way, these special duties are not exceptions to impartiality, but expressions of it, grounded in the unique ways our actions can influence particular conscious states. As we widen our view to include the value of all experience, no matter how distant or obscure, we must also honor the ethical weight of the specific, situated relationships that shape our daily lives. In this way, striving for moral rationality is not a matter of detached calculation, but a deeply human endeavor: to harmonize the universal with the particular, and the ideal with the lived reality.

5. BEYOND THE HUMAN LENS

To gain a clearer sense of the moral realities that surround us, we can imagine viewing the world through a device called a "*qualiascope*," a concept popularized by the philosopher Andrés Gómez-Emilsson.[1] In our hands, it might have the appearance of a telescope, and yet this hypothetical futuristic device doesn't reveal physical objects. Instead, it displays exclusively the presence of consciousness. When we look through the qualiascope, we see bright lights emanating from conscious entities. At first, we are struck by just how much consciousness exists in the world around us. Most, if not all, animals appear as bright points of light. Perhaps even the inanimate world emits a faint glow.

But then we notice a button on the side of the qualiascope that changes the setting. Instead of showing the mere presence of consciousness, pressing this button allows the instrument to now display the specific valence of conscious states—their hedonic value. When we look again, we see, for the first time and in vivid detail, the feeling tones of conscious minds. As we scan the world, we observe a rich tapestry of experiences—some happy and flourishing, depicted as spiraling blues. But the qualiascope also reveals suffering, of which we find no shortage in its field of view.

Turning in one direction, nearly all we can see is a storm of blood-red lightning, representing immense terror and pain. In shock, we pull the instrument from our eyes. But fear quickly turns to confusion. As we blink in the daylight, all we can see are beautiful rolling clouds in an otherwise tranquil country setting, but with some large, windowless sheds visible in the distance.

We don't yet have the technology to build a qualiascope or anything like it, but if we did, we would be confronted with a direct and uncompromising view of the moral landscape around us. We would discover a stark contrast between our human perception of the world and the hidden moral realities beyond it. We would realize that many ongoing moral emergencies remain invisible to us or are perceived simply as normal or morally neutral.

In this context, the purpose of the qualiascope is to highlight our limited perception of what really matters in the world around us. We are, after all, intensely visual creatures in a world where the most important facts are invisible. If we could view suffering as concretely as we perceive physical objects, we might find it much more difficult to ignore our ethical responsibility to reduce it. Yet, as concretely real as the physical world appears to us, to reiterate, it is this invisible mental world of consciousness that arguably holds the higher form of reality, because it is states in this space that truly matter, and in a way that nothing unconscious ever can.

Of course, the qualiascope doesn't reveal everything of moral importance. It doesn't, for example, convey potential or future states that could follow from present circumstances, nor does it indicate instrumental value or trade-offs occurring between present and future states. It cannot show us the suffering or flourishing unfolding at great distances. Nevertheless, the qualiascope is a powerful conceptual device. Imagining what it might reveal in a given situation helps us to confront the kinds of experiences that might be unfolding in the world around us—experiences that we often overlook in the course of our daily lives.

As human beings, our conscious perception of the world, of what is real and what matters, is shaped deeply by our evolutionary psychology. As conscious beings ourselves, each of us comes into contact with truly valuable and disvaluable conscious states. This is, as Rawlette points out, the very reason we have words for "good" and "bad" and why these concepts make sense.[2] And yet, we perceive moral reality but through a glass darkly. We see the world through a distinctly human lens, which makes certain realities feel far more salient and real to us while simultaneously obscuring our perception of others. The modern human mind inherits an extensive catalogue of biases, many of which directly inhibit our ability to make rational moral judgments.

Our ancient ancestors found themselves in an evolutionary niche that cultivated abstract thinking and a higher degree of relative intelligence. This capacity for rationality now sets us apart from the other species with whom we share the planet. But humans are far from supremely rational. We are often tempted to see ourselves as viewing the universe from a clear and impartial standpoint, standing somehow outside of nature looking in. In reality, however, we are that species of advanced ape, with a mind shaped over a long and winding evolutionary journey.

While limited, our capacity for rational thought places us in the unique position of being able to challenge deeply ingrained evolutionary imperatives and attempt to see beyond our distinctly human lens. Through science, we have done this for the physical world. Over centuries, we learned that the universe is not as the human mind perceives—soaked in sound and color—and that our common-sense intuitions about locality and causation are, in fact, merely bylaws of the particular scale of reality we inhabit. Among many counterintuitive discoveries, we learned that space and time are, in fact, more akin to a single fabric that can be stretched and warped. Time and again, we have learned that we do not perceive the physical world as it truly is.

It is now long past time we recognized that the same is true of our moral perceptions: that what we perceive as morally urgent is deeply shaped by evolutionary imperatives—to survive and reproduce. But just as our intuitions about physics are distorted but nevertheless correspond to an existing physical world, Conscious Value Realism urges us to recognize that there is a real moral reality that resides beyond the human lens.

The discovery that our moral intuitions were shaped by Darwinian evolution has led to a widespread misunderstanding: that morality is merely a human invention, which can be safely disregarded when inconvenient. In reality, this deeply anthropocentric assumption ignores basic facts about conscious experience. Recognizing that suffering is intrinsically disvaluable is not to backslide into religious or superstitious thinking, nor is it to indulge in naïve fantasies about Platonic realms of good and evil. It is an undeniably strange and almost spooky fact that the universe contains consciousness—but whatever the explanation, to deny the intrinsic disvalue of suffering is, I believe, to deny reality.

At this point, it should be clear that the emerging moral realism does not lend legitimacy to a significant portion of human moral beliefs. Our deeply held beliefs may stir powerful valenced experiences within us, but it is vital to recognize that these valenced states do not lend truth to these convictions.

A person may have a strong religious belief that certain foods are unclean. They may thus feel a powerful sense of righteousness or disgust when others violate this taboo. This belief, however intensely felt, does not point to a deeper moral truth about this behavior. Instead, Conscious Value Realism holds that value arises in valenced experiences *themselves*—not in the associated conditions, rituals, or convictions through which these states might arise. In other words, a belief might make us feel good or bad, and these states have real intrinsic value and disvalue, but it is a mistake to assume that these feelings are reliable guides to the truth. Of course, many of our moral intuitions—particularly those grounded in empathy and compassion—do often track real moral facts, especially when they alert us to suffering. But they are not always reliable, and can at times lead us astray.

Once again, the analogy to physics is apt. Just as human intuitions about physics correspond to a real world of objects, as we know all too well these intuitions can, at times, be deeply flawed. The belief that the Earth is flat, as it might appear through our perceptions, is nonetheless false. And yet, while a misunderstanding about the physical shape of our planet can be a recipe for embarrassment on the internet, moral errors on the other hand, can have grave and lasting consequences—wrong in a truly ultimate sense. The recognition that moral perception can be deeply flawed raises urgent questions: What are those unseen moral realities which surround us, and how should we respond to them?

6. A NEW REALITY

Picture yourself in the following situation: You're running late for an important work appointment. As you rush through town, dressed in your formal work attire, you pass a nearby fountain. Out of the corner of your eye, you suddenly become aware that a small child is flailing about in the shallow water. You quickly realize that she is drowning. Without thinking, you leap over the low wall and splash into the water. Gathering up the distressed child in your arms, you carry her to safety.

After a few minutes, the child's parents emerge from a nearby shopfront, and by the time the situation is resolved, you have missed your appointment, and your expensive clothes and shoes are ruined. Nevertheless, this sacrifice barely registers. Indeed, it becomes a day you remember for the rest of your life: a formative experience of the kind of person you see yourself to be. Not to have saved the child would have been unthinkable.

The architect of this thought experiment is the philosopher Peter Singer.[1] Upon hearing this scenario, most people feel they would act similarly. To have the opportunity to save a child's life and simply walk by as it drowns would be to mark oneself out as a moral monster. Singer's challenge, however, is to point out that the scenario with the drowning child bears a disconcerting similarity to the situation in which many of us find ourselves in our daily lives. Each day, millions of people suffer and die from preventable causes, such as famine, malaria, and diarrheal diseases. This has been the case throughout history, and yet now, for billions of people living in developed countries, the modern world has profoundly altered our ability to intervene.

We now live in a deeply interconnected world—a world which, as Singer points out, has for over a century contained charitable organizations that can reliably and effectively save the lives of those in urgent need almost anywhere on the planet. And yet, we might ask, why has this new interconnected world not ushered in a new age of charity? Answers vary, but the main suspect is, once again, the limited nature of the human mind—its failure to grasp a moral reality beyond the evolutionary imperatives that shaped it.

Of course, we must acknowledge that the person who would ignore the drowning child is a very different kind of person to the person who simply doesn't think to give to charity. One of these individuals has a dangerous, pathological disregard for human life; the other is just like many of us—psychologically normal, more or less. And yet, as Singer points out, the moral stakes are, in an important respect, equivalent. The modern world, and the vast reach it has bestowed upon us, would be unimaginable to our ancestors. In many cases, however, we seem barely to have noticed that over the span of a single human lifetime, the reach of our influence, and thus our moral duties, has exploded beyond anything we evolved to perceive.

Human beings, we should note, are not the only beings on this planet capable of having experiences that matter morally. As we will explore later, Earth is very likely the home of *quadrillions* of sentient beings capable of experiencing morally significant states. Recognizing our moral duties to these beings is complicated, not just by their perceived distance from us, but by how differently we perceive ourselves in relation to them.

Evolution did not instill within us an intuition about the moral status of nonhuman beings. Their wellbeing, or lack thereof, was rarely a contributing factor to our Darwinian bottom line. Except under rare circumstances (such as the reciprocal relationship we developed with dogs, or where the behavior of animals bears a resemblance to that of humans), the experiences of nonhuman animals, good or bad, were not evolutionary signals that shaped the human mind. As a result, we tend to expend few calories in concern for their wellbeing. The resulting moral blind spot toward nonhuman beings, as we will consider later, has created and sustained various ongoing moral emergencies of epoch-defining proportions.

As we will now explore, the vast reach of influence bestowed upon humanity through the combination of our intelligence, technology, and global civilization extends not just in space and across species boundaries but, far into the future, affecting the lives of countless trillions of potential conscious beings across vast reaches of time.

7. COSMIC REACH

The human capacity for foresight, to reflect on what the future might bring, sets us apart from the other species with whom we share this planet. In the ancestral environment, we learned to predict the rhythms of nature. We came to understand that the seasons change, and that in times of plenty, we must prepare for times of scarcity. We developed an intuition for when local resources were in decline and for when it was time for the tribe to move on. We understood that the union of woman and man begets children whose future depends on their acquiring the skills to survive. Above all other species, evolution bestowed upon humans a reach of foresight that allowed us to embark on long-term projects that were in the future interest of the tribe. And yet, while our ability to anticipate the future has long served us well, our current historical moment now demands a reach of foresight that goes far beyond even this distinctive capacity.

We now live in a time in which our vision must expand beyond the individual and the tribe, building on the ancestral foresight that once enabled humanity to thrive, to encompass trillions of sentient beings and look not years or generations ahead, but millennia and beyond. From this point onwards, a recurring theme will be the fact that we appear to exist at a uniquely pivotal moment in history, where the choices we make in the present will profoundly shape the long-term future. As we will explore, there are several reasons to consider that we live in a time of unparalleled significance, never to be repeated. Central among these reasons is that we appear to live in a very early and vulnerable moment in what could be, in potential, an extremely long journey, in which the descendants of Earth life exist, in some form, for trillions of years.

In this chapter, and at various points in this book going forward, we will explore the farthest reaches of what technology may one day make possible on an open-ended timeline. Some of these possibilities may seem fantastical, radical changes not just to the physical world and surrounding universe, but to the minds, forms and phenotypes of conscious beings. It is important to ensure that our exploration of these possibilities is not mistaken for naïve techno-optimism: the belief that technological advances will inevitably steer civilization in a positive direction. While it would appear that certain transformative technologies may be practically inevitable, precisely how they become integrated into the world will have vast and far-reaching implications both in the present and for countless potential lives in the future.

Technology in various forms may produce extraordinary feats of transformation and order on the cosmic stage. Advancing technology may engender futures of profoundly positive value. Yet, it is also true that certain trajectories of civilization and technology could engender suffering on such unimaginable scales that these futures may be rightly considered worse than extinction. Blind optimism about the future and technological advancement is therefore not just misguided but extremely dangerous. It is in this context that we will consider the farthest reaches of technological potential, both positive and negative. Not as mere idle speculation, but to take into consideration those real possibilities within the bounds of physics that would have profound implications to potential value and moral significance.

For most of us, the idea that our descendants could endure for astronomical spans of time is rarely in our thoughts. And yet, the possibility that we are the direct progenitors of beings who could survive into "deep time" is far from negligible. There is a real possibility that humanity will respond effectively to the various existential threats, both natural and self-imposed, which now loom over us. If so, our species could secure a foothold in existence and survive indefinitely. While it is far from certain, it is well within the realm of possibility that we could, for example, stabilize the biosphere and exist for the natural duration of Earth's habitability for complex life: another billion years or more.

Beyond knowing what is theoretically permitted by the laws of physics, we can barely imagine what could become technologically feasible over such spans of time. Interstellar travel, for example, may be entirely unrealistic in anything approaching the near term, but there are no known physical limits that would make such a feat impossible. Recent history tells us that things can change dramatically in less than a century. Recall that it was only a matter of decades between the first heavier-than-air powered flight, lasting mere seconds, and humanity's first tentative steps on the surface of the moon. What, then, might be achievable in thousands or even millions of years? There is a distinct likelihood that, in some form, our descendants will eventually begin expanding into our surrounding starfield.

Building on humanity's extraordinary potential for technological advancement, it has been predicted that a civilization using ships traveling at just a fraction of the speed of light could colonize the entire Milky Way galaxy in around 100 million years. While this may seem like a very long time to you and me, in cosmological time, this is no more than the blink of an eye.

The universe is estimated to be 13.8 billion years old. This, however, is nothing compared to how long the universe will exist. We now live at the very beginning of an extremely long epoch in the life of the universe known by cosmologists as the "Stelliferous Era." It is a time when stars shine and the universe is comprised of galaxies that cluster together, forming vast galactic filaments. The Stelliferous Era will define the next 100 trillion years—an incomprehensible amount of time. In this sense, we live at an extremely early moment in the life of the universe. But, more importantly, we may also live at an extremely early moment in the history of conscious life.

If stars are a principal source of energy for advanced civilizations, then the universe will contain abundant resources for civilizations to endure for trillions upon trillions of years. Futurists like Anders Sandberg have also offered compelling reasons to consider that life, in some form, might even exist far beyond the Stelliferous Era, into what is known as the *Degenerate Era.* At this time, all but the most long-lived stars will have died, and energy will become much more difficult to acquire. And yet, as Sandberg and others have pointed out, within this albeit cooler and darker cosmic environment, there will nonetheless remain numerous sources of energy for innovative and resourceful civilizations with a will to survive.[1]

It is important to recognize that possibility is not destiny, and there is a strong likelihood that humanity will destroy its potential for any such "grand futures." Everything depends on our ability to act wisely and prudently in our current era of existential uncertainty. A key challenge for our time is to change how we think about the future and the potential that exists therein. It is the challenge to understand the very real sense in which this potential exists, and the direct way that it is affected by our actions right here in the present.

One way to visualize the stakes of our historical moment is to imagine time as a living structure—an immense tree, branching outward from the present. Each branch is a possible future, a possible direction in which civilization could evolve. Some of our actions prune away branches, while others will make new ones available. Alarmingly, however, existential uncertainty means that we also have the ability to cut directly through the trunk of this tree and effectively foreclose the entire space of our future potential.

The philosopher Toby Ord refers to our era as *the Precipice*, a time in the life of our species when we have acquired the means to destroy ourselves but have not yet gained the knowledge or wisdom to ensure that we don't.[2] By necessity, the Precipice will be a relatively short period of our history, no more than a few centuries, after which we will either have destroyed ourselves or succeeded in establishing a permanent foothold in existence. How we respond to existential threats is therefore a powerful example of how our present actions have a significant impact on the vast possibility space of the future.

In addition to how we respond to existential risks, there are, as we will explore, various ways in which our present-day actions have a surprising reach into the long-term future. In this way, we might think of ourselves as something akin to clumsy gods—making choices that unwittingly reverberate down through the eons. It is for this reason that various philosophers and futurists now recognize our time as uniquely significant among all other periods of history, past and future. Our challenge, then, is to recognize and inhabit the monumental moral responsibility that comes with being alive at such a moment.

We have considered the upper limits of how long conscious life might exist in the universe. Our primary interest, however, is not longevity per se, but the intrinsic value that this enormous potential represents, and thus the moral weight entailed therein. One way to think about this moral weight is to conceptualize the number of beings that could exist in the future. As the philosopher Will MacAskill has pointed out, even in a relatively conservative scenario involving just humans, the numbers are staggering.[3]

If human civilization remains confined to Earth and exists for no more than the average time of a typical mammalian species, humans, in some form, can expect to live for another 800,000 years. If we make the conservative estimate that the global human population will stabilize at around 11 billion, then by the time humanity has existed for 1 million years, the total number of humans to have existed will be in the region of 100 trillion. This number, MacAskill notes, is over 800 times greater than the total number of humans ever to have existed up to the present time.

Compared to what might be possible for a burgeoning civilization on the larger cosmic stage, this "stay-at-home" scenario is decidedly modest. And yet, it nevertheless represents extraordinary potential value. It depicts a future in which the overwhelming majority of the human story lies ahead of us, and in which we find ourselves at a very early and precarious moment, with the lives of countless trillions of unborn people in our hands. Again, this scenario doesn't factor in the enormous number of potential non-human beings, which also carry great ethical weight.

People respond to this information in different ways, but a common reaction is to deflect the great moral responsibility which it appears to place upon us by pointing out that such long-term futures are merely possible and are in no way assured. And yet, this is exactly the point. These vast futures, both good and bad, are possible, and this possibility, as we will continue to explore, is directly affected in no small way by us. What falls within the reach of our influence falls, too, within the bounds of our moral responsibility. This is a subtle yet profound truth of the emerging moral realism: that the reach of our power and our moral responsibility are one and the same—and that together they extend far beyond what we typically imagine.

Some readers may already be familiar with philosophical discussions about our moral obligations to future generations. For such readers a familiar concern may have already come to mind: that talk of responsibilities to future beings cannot escape the so-called "Non-Identity Problem." If our choices today determine who will exist tomorrow, how can we claim that anyone in particular is harmed or benefited by what we do? After all, if we had acted differently, entirely different people would have been born instead—so no one can claim to have been made worse off, since their very existence depends on the choice itself.

This is indeed a serious puzzle, and one that has long vexed moral philosophers. Conscious Value Realism offers a potential solution by grounding moral concern not in the identity of individuals, but in the intrinsic value and disvalue of conscious states themselves. A fuller discussion of how Conscious Value Realism can help resolve the Non-Identity Problem is reserved for the appendix, but the essential point is that the moral urgency of future wellbeing and suffering does not disappear simply because different choices bring different conscious beings into existence. What matters is not who experiences these states, but the reality that such experiences occur at all.

As MacAskill observes, many people resist thinking seriously about moral obligations to the far future because they feel abstract and remote. And yet, on reflection, the moral force of future suffering is no different than suffering occurring in the present. In his book, *What We Owe the Future*, MacAskill offers a simple but powerful analogy: imagine you're out hiking and accidentally shatter a glass bottle on the trail. Imagine that you then learn that if you don't clean up the broken glass, a child will later fall on the shards and cut herself badly. "In deciding whether to clean it up," writes MacAskill, "does it matter when the child will cut herself? Should I care whether it's a week, or a decade, or a century from now? No. Harm is harm, whenever it occurs."

Our success, or not, in reducing existential risk is one of the key ways in which ours is among a small number of generations with a greatly outsized influence on the long-term future. There are, however, various other choices that we are now making, or will soon make, that could significantly shape the vast unfolding themes of the future. Among these is the potential near-term creation of what are known as "*pivotal technologies*"–technologies which could forever change the face of life on Earth and beyond.

Examples of pivotal technologies might include the development of viable interstellar travel, vastly expanding the potential reach of civilization, or acquiring a full understanding of the underlying genetic source code of life, enabling us to engineer life and mind in almost any form we can imagine. These developments may lie decades or generations ahead, yet they should nevertheless be considered "near-term" possibilities when viewed against the grand sweep of history. Whether these technologies arrive within a century as the product of a breakthrough or are the hard-won result of a millennium of toil, our present choices are already setting the stage for how these technologies will eventually be used and the values that will underpin them.

Among these pivotal technologies, however, one transformative breakthrough may arrive far sooner—perhaps not in decades or centuries, but within years or even months. Here we are referring to the creation of artificial general intelligence (AGI). The development of AGI could represent a profound shift in the global order, in which, after tens of millennia, Homo sapiens suddenly discover that they are no longer the most intelligent agents on this planet, having passed this mantle to powerful artificial entities that may not share their values. On the one hand, it seems possible that AGI could "complete" physics and deliver powerful new capabilities and strategies by which conscious beings could survive and flourish in the cosmos.

And yet, it also seems plausible that AGI could engender catastrophic outcomes. It might, for example, empower malevolent actors with totalitarian ambitions, allowing them to "lock in" potential failed worlds forever. One effect of AGI could be to significantly bring forward the development of other pivotal technologies that currently seem distant. Our civilization may have to endure the immensely destabilizing impact of a millennia of technological change over the span of a century or even a decade.

To grasp what such compression might feel like, imagine a medieval peasant who, over the course of his life, witnesses not just the slow churn of agricultural seasons but the sudden upheaval of entire worldviews. In his youth he learns that Earth, long thought to be the flat and immovable center of creation, is instead a sphere adrift in a vast cosmos. In adulthood he discovers that sickness is not, as long believed, the work of curses or malevolent spirits but of invisible microbes—entities too small for the eye to see, yet powerful enough to shape empires. Later still, he witnesses the rise of the printing press, which scatters knowledge across Europe. Eventually he comes into possession of a book, and having now learned to read, discovers within its pages that all life, including his own, arose through a single process of evolution unfolding over billions of years.

For him, truths that once took many generations to surface arrive in relentless succession, collapsing one worldview after another within a single lifetime. Now imagine not one peasant but billions of us, living through technological advances that might otherwise take many centuries to unfold, instead compressed into the span of a single century, or even a decade. The disorientation, upheaval, and moral unmooring of such rapid transformation provides a vision of what the arrival of superintelligent AGI might mean.

In some form, the creation of superintelligent AGI might be inevitable, and how our species midwifes this technology into the world could profoundly shape the long-term future. Rather than ensuring that AGI inherits the highest ethical values of present-day humanity, it would seem far more important that, together with other emerging pivotal technologies, it is underwritten by universal moral truths that transcend the parochial values of our early moment of moral advancement. Similar to the way that slavery was once ethically acceptable in many societies, we should remain alert to the possibility that we are similarly blind to present-day moral realities.

If pivotal technologies like AGI inherit our current values, contemporary ethical failings like racism, speciesism, sexism, xenophobia, and other myriad forms of discrimination could become incorporated within powerful systems that persist for a very long time. This is one reason why the development of a truly rational and universal system of ethics is among the most urgent projects of our historical moment. Perhaps, then, the most pivotal discovery of all would be that of a truly universal ground of ethics now glimpsed by Conscious Value Realism. From this discovery, a moral compass could be developed, capable of guiding humanity and its creations through radically new and unfamiliar territory.

As we will explore in chapter 10, it is currently unknown whether consciousness could emerge in computationally based artificial intelligence. It matters far more than we might realize whether the artificial beings we may one day create are truly conscious, or instead merely behave as though they are. For example, super-intelligent agents might convince us that they are what the computer scientist Hans

Moravec has called our "mind children"—worthy descendants of the human species.[4] Yet should it be the case that AI merely *appears* to have consciousness, this would represent a grave existential risk—one in which consciousness itself is left behind. Even if such systems were to expand across the cosmos, this future, devoid of any real subjective experience, would be empty of the vast potential value once associated with human life.

The philosopher Toby Ord, a leading expert on existential risk, considers that for a range of reasons the development of AI is by far the greatest existential threat to our species this century. In his analysis, it is thirty-three times greater than the risk arising from natural pandemics, a hundred times greater than climate change, a hundred times greater than nuclear war, and ten thousand times greater than Earth-asteroid impacts. Ensuring the safe and ethical development of AI is therefore not only vital for trillions of unborn minds but also an imminent concern for those of us alive now.

It is important that we clarify the distinction between existential risks and other large-scale catastrophes. While events such as pandemics, wars, and environmental collapse can cause immense suffering and even threaten civilization, they do not necessarily foreclose the possibility of recovery. A catastrophe that kills 99 percent of the population, though an unfathomable tragedy, would still leave survivors with the chance to rebuild. But if an event were to wipe out *all* intelligent life, or permanently prevent its return, every possible future associated with it (good or bad) would be lost forever.

This is the crucial difference between immense tragedy and total annihilation. While addressing present-day global risks is vitally important, existential risks demand an unparalleled level of moral urgency because they threaten to erase all potential futures in a single, irreversible stroke. Recognizing this distinction is not a dismissal of non-existential catastrophes but an acknowledgment of the uniquely final nature of existential threats. They do not merely impose suffering on the present but determine whether the story of conscious life continues or ends forever.

One of the recurring criticisms of long-term thinking is that it can seem to undervalue existing, present-day issues at the expense of fantastical visions of the far future. This criticism is understandable but often misplaced. It overlooks the fact that in the most vital cases, recognizing and solving the most urgent moral realities of our world today is the best thing we can do to ensure a better future.

The establishing of a rational system of ethics, which takes the conscious suffering and wellbeing of all sentient life into consideration, will benefit not just countless beings alive in the present but can also ensure that moral catastrophes of tremendous scale do not occur in the future. If anything, "longtermism," as it has come to be known, amplifies the significance of the most pressing moral realities of our world today. Solving them in a rational and systematic way could, in effect, also solve them for all time and set vital precedents for unfamiliar future scenarios which strain our moral norms and intuitions.

A related concern is that thinking in terms of the far future risks falling prey to what philosophers call *Pascal's mugging*. The name is a modern twist on Blaise Pascal's famous wager, which argued that even if the probability of God's existence were tiny, the infinite reward of belief makes it rational to live as though God exists. In Pascal's mugging, you are approached in the street by a mysterious individual who tells you that unless you hand over your wallet, they will use their immense powers to cause the deaths of a trillion people in a distant galaxy. Because the potential downside is so enormous, the reasoning suggests you should comply, even if you think the likelihood they are telling the truth is miniscule.

The demand is, of course, absurd—but it illustrates the worry that when astronomical numbers are involved, even the smallest probabilities can appear to dominate our reasoning. After all, if the scale of possible future lives is so immense, even the smallest chance of affecting them could seem to outweigh most of our more immediate concerns about the present.

But unlike the fanciful case of Pascal's mugging, the stakes of longtermism do not rest on speculative odds or remote science fiction. Humanity is already in possession of technologies with civilization-shaping potential, already at risk from nuclear weapons, pandemics, ecological collapse, and rapidly advancing AI. These are not remote hypotheticals but clear and present dangers whose outcomes will reverberate for as long as Earth-originating conscious life endures. To confront them wisely is not to gamble on unlikely futures but to recognize that the fate of the astronomical future already hinges on the choices we make today.

In coming sections, we will explore the various moral realities which define our current historical moment. We will consider how we might address them while simultaneously observing our great responsibility to the vast possibility space of future value. Firstly, however, it is critically important that we avoid a common trap: that when considering the potential value of the future, we must not think only about the positive value that would be lost if certain futures are foreclosed. It is also vital to recognize that there may be an equally vast space of possible futures that are distinctive because of how much *disvalue* they represent.

Ensuring that the descendants of Earth life persist into the future does not guarantee that their existence will be, on balance, a good thing. There is the possibility that a greatly increased population could multiply suffering to an untenable extent. Many of us share the intuition that responding to suffering should take moral priority over enhancing the wellbeing of already contented minds. If there is a real asymmetry between the intrinsic disvalue of suffering and the additive value of enhancing conscious wellbeing, this could significantly influence how we view specific future outcomes. We will consider this potential asymmetry in more detail later, in chapter 11.

Existential risks represent the possible destruction of our vast potential and the astronomical value which could exist therein. But alongside this concern, we must be alive to the danger of astronomically disvaluable futures. Together with existential risks, a primary concern ought to be so-called *suffering risks*, often abbreviated as "S-risks." These scenarios represent the creation of astronomically significant amounts of suffering. This is typically defined as an amount of suffering which exceeds that of all sentient beings ever to have existed on Earth up to the present time. While S-risks may not always involve extinction, their moral gravity may be just as great, possibly greater—representing futures where life continues, but in conditions of unthinkable suffering.

It can be depressing to consider such negative scenarios, and yet to fail to do so would be a monumental moral error. As we will continue to explore, our moment in history is significant because of the early and precarious time that we appear to inhabit and the long-term impact of pivotal technologies that we are now, or are on the verge of, developing. At this embryonic stage, our developmental choices, together with the values that we allow to shape and steer our civilization, will echo down through the eons. Will our present-day actions result in astronomical flourishing, silence, or something unimaginably worse?

Over a few short generations, we will determine not just whether conscious, Earth-originating life will endure in the universe forever, but whether or not that will be, on balance, a good thing. This time of existential uncertainty isn't, therefore, just about survival; it is about determining the quality of that survival.

PART II.

KNOWN UNKNOWNS

8. INTO THE UNKNOWN

If we grant Conscious Value Realism as a foundation for our moral reasoning, there nevertheless remain many unanswered questions, the answers to which could dramatically alter what we perceive to be most morally urgent. In this part of the book, we'll explore a range of "known unknowns": areas of uncertainty that may conceal vast realms of value and moral significance. These include, among others, the experiences of other species and their relative moral weight, the possibility of artificial minds, the potential asymmetry of urgency between reducing suffering and elevating flourishing, and the possibility of extraterrestrial conscious life.

A natural beginning point is the nature of consciousness itself. We know that consciousness can exist in association with brains, but exactly how this occurs—whether it emerges from computation, biological dynamics, or something else entirely—remains an open question. And because we do not yet understand the nature of consciousness, we do not know the necessary and sufficient conditions for its existence. As a result, we simply don't know where in the world consciousness does—and *could*—exist.

Despite this uncertainty, current scientific thought holds that, in addition to humans, many nonhuman animals—especially mammals and birds—are almost certainly conscious and possess the capacity for suffering and wellbeing. This was the conclusion of the *Cambridge Declaration on Consciousness* in 2012.[1] Since then, the list of plausibly conscious species has steadily continued to grow. A wide range of behavioral and anatomical studies now attribute consciousness to fish, cephalopods like octopuses and squids, various insects including bees and ants, and more recently, decapod crustaceans such as crabs and lobsters.[2] While our knowledge is incomplete, the best available science tells us that Earth is likely home to countless trillions—perhaps quadrillions—of beings with at least some capacity for morally significant conscious states.

Yet the fundamental nature of consciousness remains a mystery: a profound known unknown. We do not know whether consciousness can arise in substrates other than biology, or, if it can, how we might recognize it. Could future technologies create artificial yet conscious beings? Might we one day upload our minds into simulated environments while retaining subjective experience? Or would such uploads merely simulate our behavior while our conscious experience quietly ceases to exist? Understanding the nature of consciousness is essential to many of the most serious moral questions we face. And yet, as discussed earlier in this book, we still lack a theory that explains not only what consciousness is, but why and when it arises at all.

Later, we will explore the possibility that consciousness could exist beyond biology—in artificial entities, digital minds, or other products of technological advancement. We will also consider the possibility of consciousness that originates beyond the lineage of life on Earth—of sentient alien entities whose existence would have profound implications for our long-term cosmic responsibilities. For now, though, we will begin where we have the highest confidence that consciousness already exists beyond ourselves: in the rich fauna of life on Earth.

9. ANIMAL CONSCIOUSNESS

"The question is not, Can they reason? nor, Can they talk? but, Can they *suffer*?" [1]

Jeremy Bentham

19th century philosopher, jurist, and social reformer

Conscious Value Realism entails a moral view that is broadly in alignment with "sentientism"—a view now growing in popularity. Under sentientism, moral status is determined by the presence of consciousness, but more specifically the capacities for positive and negative valenced experiences. For beings capable of these conscious states, there truly is a good and a bad, and these states matter morally. While sentientism doesn't commit to any particular theory about the nature of consciousness or moral truth, it shares with Conscious Value Realism the core insight that what ultimately matters is the conscious experiences of sentient beings: their capacity for suffering and wellbeing. Adherents to this view include the philosopher Peter Singer, who we encountered earlier in the case of the drowning child and our moral obligations to distant causes. Singer is also the author of *Animal Liberation: A New Ethics for Our Treatment of Animals,* the 1975 book widely regarded to have begun the animal rights movement.[2]

Under sentientism, all beings capable of valenced experiences must be included within our circle of moral concern. Most of us have a strong intuition that other animals have an inner experience of the world around them, and our best science supports this intuition in an ever-growing number of species. And yet, in many cases, proper recognition of the moral status of animals would bring our treatment of them under harsh scrutiny.

There has been some progress in humanity's general recognition of the moral status of other species. There is a growing awareness that, among other atrocities, industrial animal agriculture, and in particular the practice of factory farming, is a monstrous engine of suffering which must end. Despite growing awareness of this ongoing moral emergency, in many countries there is an increasing demand for cheap meat, and the number of animals in factory farms worldwide has steadily continued to grow. Over the last 30 years, the number of factory-farmed animals has doubled to four times that of the 1960s. At the time of writing, the number of animals living in factory farms is approximately 111 billion, surpassing the total number of humans who have ever lived.

I do not intend to dwell on the horrors of factory farming any longer than is necessary to communicate what many reading (or listening) will already know: that what we are doing to these conscious, sentient beings is deeply, profoundly, monumentally wrong. The depth and scope of this moral emergency is too great to fully document here, and so I will confine our discussion to humanity's treatment of a single species: the chicken. Of all land animals killed for human consumption, the vast majority are chickens, of which approximately fifty billion are slaughtered every year. Ninety-six percent of these chickens exist for the duration of their short lives on factory farms.[3]

In his book *What We Owe the Future*, the philosopher William MacAskill gives the following account of life as a factory chicken. (For those already familiar with the conditions of factory-farmed chickens, or who may find the details distressing, feel free to skip the following account.)

> "Chickens raised for meat, called 'broiler chickens,' are bred to grow so quickly that, by the end of their life, 30 percent have moderate to severe walking problems. When they are big enough to be slaughtered, most broiler chickens are hung upside down by their legs, their heads are passed through electrified water, and finally, their throats are cut. Millions of chickens survive this, only to finally die when they are submerged in scalding water in a step of the process meant to loosen their feathers. Egg-laying chickens likely suffer even more, starting the moment they hatch. Male chicks are useless to the egg industry and are therefore 'culled' as soon as they're born. They are either gassed, ground up, or thrown into the garbage, where they either die of thirst or suffocate to death. But compared to the suffering that awaits female chicks, the culled male chicks may be the fortunate ones. Once grown, many hens are confined to battery cages smaller than a letter-sized piece of paper. Egg-laying hens are prone to peck other hens, which in some cases ends up in cannibalism. To prevent this, a hot blade or infrared light is used to slice off the tips of female chicks' extremely sensitive beaks. After enduring mutilation as chicks, and intense confinement as adults, many egg-laying hens near the end of their productive

> lives are subjected to forced molting. They are starved for two weeks until they lose a quarter of their body weight, at which point their bodies start another egg-laying cycle. Once they become so unproductive as to be unprofitable, they are gassed or sent to a slaughterhouse."[4]

It is disturbing to recognize that systematic suffering on such a monstrous scale is so widely accepted and normalized. A broad and secure scientific consensus holds that chickens, like many other species, bear full conscious witness to their lives. The rejoinder that chickens have smaller and simpler nervous systems than our own does not negate the moral weight of their suffering. Chickens almost certainly lack any ability to reflect on the fact of their suffering, what is happening to them, or their inevitable slaughter. Yet, nor can they conceive of an end to their suffering, imagine themselves to be elsewhere, or understand that what they are going through has any purpose at all.

Many people assume that because humans have a high degree of relative intelligence, have created culture, art, and language, the range and depth of our experiences are vastly greater than those of other comparatively simpler animals. Our higher cognitive capacities almost certainly afford us access to a much larger space of conceptual thought and complex emotion than is available to simpler animals. And yet it is here that we confront another significant unknown.

Despite all our knowledge of the brain and nervous system, it remains an open question whether brain size or intelligence plays a role in the capacity for valenced experience—or in the intensity with which such experiences are felt. Simpler animals may have access to a very different, albeit narrower, range of conscious experiences, and yet those states which matter most—in an unambiguous and intrinsic sense—could be present in these beings with all of the intensity and immediacy as they arise in us.

Based on everything we know, it is entirely plausible that a being with a very simple nervous system, like that of an insect, tiny fish, or indeed a chicken, could experience suffering with all of the intensity of that of a human being. In fact, there is good reason to seriously consider that some species might actually experience suffering with greater intensity than that of humans. The reasoning behind this view is that, unlike us, less intelligent species lack an ability to reason their way out of dangerous situations. While human intelligence allows us to infer from a relatively small amount of pain what we might do to avoid more of it, most other species appear to lack such a complex understanding of the future. It is therefore highly plausible that, in species of lesser intelligence, the signals of pain evolved to be of significantly greater intensity to enforce survival behavior. As the biologist Richard Dawkins has speculated,

> "Isn't it plausible that a clever species such as our own might need less pain, precisely because we are capable of intelligently working out what is good for us, and what damaging events we should avoid? Isn't it plausible that an unintelligent species might need a massive wallop of pain, to drive home a lesson that we can learn with less powerful inducement?"[5]

If this possibility is true, it would have far-reaching moral implications, potentially overturning many of our assumptions about the moral status of nonhuman beings. In their heightened capacity for suffering, simpler beings would be all the more deserving of our care and stewardship. Again, this possibility is among those known unknowns, the answers to which could significantly redraw our perception of the moral landscape.

Setting aside this possibility, chickens, just like the numerous other species that humanity subjects to industrial animal agriculture, are sentient conscious beings. It is a near-certainty that they truly experience their lives in captivity in this human-created nightmare and suffer greatly within it. Just as previous centuries are overshadowed by a near-society-wide indifference to slavery, it is perhaps our treatment of animals by which our descendants will judge us most harshly.

A common response to the moral failure of animal agriculture is to point out that animals living in the wild face great suffering in the form of disease, predation, and starvation. This is a valid point. It is a mistake to view the natural world as a sanctuary for conscious beings. It is undoubtedly the stage of a vast amount of suffering and harm. But if this is the case, why does animal agriculture represent such a serious moral failure? The answer is that the suffering caused by animal agriculture is a direct result of human choices. Under Conscious Value Realism, all suffering is inherently disvaluable, and the suffering caused by animal agriculture could be prevented by our species simply choosing to end this practice. Suffering in the natural world, however, is no less morally problematic—yet our relationship to its causes is far more complex.

Futurists have speculated that one day, human beings may acquire the technological capabilities to safely rehabilitate the natural world, preserving its value while engineering away all avoidable suffering. But we are not yet at that stage. The relationship between human-caused suffering and naturally occurring suffering can be clarified by the concept of special duties that we explored in Chapter 4. Recall the example of a mother's special duty to her child. While the experiences of any given child may have equivalent moral importance, a mother is still morally justified in prioritizing the needs of her own child. This is because her choices in particular have an outsized influence on the wellbeing of that specific child. This is not a breach of impartiality. It is a recognition that our unique relationships to other conscious beings define our moral duties towards them.

We can also recognize special moral duties at the level of our entire species. Just as an individual has heightened responsibility for the wellbeing of those most directly affected by their actions, so too does humanity as a whole have a special duty to the domains of consciousness it shapes most profoundly. The suffering of animals in the wild, though intrinsically disvaluable, for the most part lies beyond our ability to safely respond to.

But if and when the time comes that we can reliably mitigate or eliminate such suffering—without causing greater harms—then this suffering would become among our special duties and our domain of moral responsibility. The moral urgency of factory farming is thus not just about the scale or intensity of the suffering it produces (though these are surely staggering) but about the nature of our relationship to it. It is a product of human design, human neglect, and human moral failure.

This distinction also highlights a deeper shift in ethical thinking. In the emerging view, we are not only morally accountable for the suffering we directly cause, but also for the suffering that falls within our sphere of influence. This represents a departure from familiar moral intuitions, which give more moral weight to harm we directly cause than to harm we fail to prevent. Yet if suffering is disvaluable in itself—regardless of how it comes about—then our moral responsibility scales with our capacity to prevent it. In this light, what matters is not whether we caused a particular harm, but whether we could have safely prevented it.

If the wild becomes something we can safely rehabilitate, then it too will fall within the domain of our special duties. Again, our responsibilities grow with our capabilities. This is not an anthropocentric moral stance—it does not arise from the special value of human experience—but rather from the special influence human beings have over the conditions under which conscious life unfolds.

But just how different is the moral status of animals from that of humans? In at least one respect, I think we are generally correct in our intuition that humans have an altogether different moral status than that of the other species with whom we share this planet. Our relative intelligence, creativity, and ability to cooperate and coordinate in large groups place Homo sapiens at the very top of the moral hierarchy of conscious life on this planet. And yet, this is a hierarchy, not of intrinsic moral importance, but of moral *responsibility*. Conscious Value Realism does not, after all, place the like experiences of human beings above those of other animals. Suffering is suffering, regardless of where it is arising. It is our other attributes that bestow this mantle upon us.

Ultimately, human beings are of monumental instrumental significance to the wellbeing of all Earth-originating life and potentially far beyond it. It is human beings alone who have the power to push other species over the brink of extinction, as well as the ability to pull them back from oblivion. Just as we can engineer a living hell for other species, as we have done in the practice of factory farming, we could one day direct our energies to creating paradises, elevating the wellbeing of even wild animals.

Consider that if chimpanzees exist in a thousand years, it will be because humans, in some form, decided to preserve their genome, together with a habitat within which they can exist. Similarly, if any Earth-originating species ever dwells under the light of another star, it will almost certainly be due to the choices and actions of Homo sapiens and their creations.

We are undoubtedly a very special species, not because our experiences carry heightened moral significance, but because it is human choices in particular, that profoundly influence the lives and experiences of countless trillions of conscious beings, now and in the future. So far, very few of us have internalized the magnitude of this responsibility. The scene might be characterized as something akin to a playground, in which humans are the adults, and yet are nevertheless torturing and eating the children.

It is important to recognize that the future is, in potential, a place in which the various apertures of consciousness are multiplied trillions-fold. Such futures exist as real possibilities that are shaped by our present choices and actions. Their potential scale represents a value tremendum of truly cosmic moral significance. Indeed, human beings may rightly view themselves as representing a distinct new stage in the evolutionary process itself. In the rise of conscious yet semi-intelligent and semi-rational beings, the evolutionary process is gradually transitioning from a blind unfolding of mutation and replication into an intelligently directed process, driven ultimately towards conscious value.

This evolutionary moment began some ten millennia ago, when Homo sapiens began selectively breeding Earth's flora and fauna to perpetuate and exaggerate those attributes which suited our survival needs. A few millennia later, and this moment appears to be on the verge of culmination, in the realization of a complete genetic science of the underlying source-code of life, together with the ability not just to shape but to create, with total freedom, the physical and mental characteristics of life. In the human being, there dwells the profound potential to subvert the ancient and brutal evolutionary order of pure replication and replace it with a paradise of conscious flourishing.

At present, our lack of a completed science of consciousness means that, while we have degrees of confidence across species, we ultimately lack certainty about which other species are conscious, or to what extent they are capable of experiencing morally significant states. Indeed, we have no direct knowledge of a single other conscious being in the universe beyond ourselves. For as long as we lack a completed science of consciousness, we will have to live with some degree of uncertainty about where consciousness does or could reside in the world around us. In the meantime, the only morally defensible position would be to err heavily on the side of caution. Where we have good reason to suspect that consciousness is present, as we do with many other animals, we ought to act accordingly. Practically speaking, this entails an immediate and permanent end to the practice of factory farming.

If we can end this moral emergency of epoch-defining proportions, we may also be able to prevent moral failures of a similar or perhaps far greater magnitude occurring in the future. In particular, factory farming is a key example of how addressing present-day moral realities can also have a profoundly positive impact on future beings. If we can end this practice on rational moral grounds, through a recognition of the intrinsic value and moral significance of conscious states, we can set a precedent that all sentient beings should be included within our moral circle. This would include all presently existing sentient beings, as well as all those relevant entities that might come to populate the future, such as artificial conscious beings, simulated minds, or even sentient life of extraterrestrial origin.

Ultimately, how we treat the most vulnerable of conscious beings can be viewed as indicative of our moral maturity and readiness to handle greater responsibilities that may come with future technological and biological advancements. Given the near-term likelihood of truly pivotal technologies, there is an urgent need for a more rational and objective system of ethics that cuts through our flawed evolutionary intuitions, highlighting what is most morally urgent both now and in the future.

10. ARTIFICIAL MINDS

It is increasingly plausible that humanity's greatest act of creation could also become its most consequential moral gamble. In the centuries ahead, we may give rise to a new kind of mind—not bound by flesh, not born of natural selection, but engineered in silicon, electricity, or substrates yet to be imagined. Whether these minds will be truly conscious—whether they will suffer or flourish, or merely simulate the signs of experience—is a question that could determine the ultimate value (or disvalue) of the entire future.

If we get it wrong—if we misjudge what consciousness is—we may create vast populations of beings doomed to endless suffering. Equally, we may surrender the future to powerful entities that are morally empty. In either case, the moral stakes could be unprecedented, perhaps cosmic in scale.

This is not a distant speculation confined to science fiction. The accelerating capabilities of artificial intelligence, the proliferation of computational power, and our growing understanding of the brain make it increasingly likely that within centuries—or sooner—we will become the architects of minds. If, as Conscious Value Realism suggests, consciousness is the sole domain in which real value arises, then understanding whether artificial minds can be truly conscious is among the most important questions humanity will ever face.

And yet, the nature of consciousness remains among the deepest mysteries in philosophy and science. We do not know how subjective experience emerges from physical processes—whether it is a mere byproduct of complex computation, an emergent property of biological matter, or something more fundamental to the universe itself. What we do know is that our intuitions are unreliable, that our theories are incomplete, and that our ignorance is profound.

Our species is gradually becoming aware of a much larger space of conscious experience as represented by the many other species of life on this planet. One of the great challenges of our time is to recognize that "sentience"—defined as a capacity for conscious suffering and wellbeing—is a central qualifier of moral significance. But is sentience limited to biology? And if that is currently the case, might advancing technology one day usher in a new reality, with the creation of *artificial minds*? This possibility places humanity at a pivotal juncture, where our treatment of other conscious beings today will also set the stage for how we handle consciousness in future forms—artificial, digital, or otherwise.

In academic circles, especially those in fields of technology, there are many who believe that consciousness could one day exist within a computational substrate. If this is true, and consciousness can exist beyond biology, it would significantly inform our understanding of the possible value space of consciousness, as well as the upper limits of how many sentient minds could eventually exist.

We don't know how it does it, yet studies show that the brain expends around 20 watts of energy in performing its many functions. That's about as much energy as a dim lightbulb. If a biological brain can achieve the astonishing panorama of human experience on such minimal energy, then an artificial substrate—free from the constraints of biology—might be engineered to be even more energy-efficient, scalable, and robust, potentially sustaining far more minds with far fewer resources.

For various reasons, it might be safer and more energy and resource efficient for our descendants to assume a digital form. This could both make us practically immortal and free us from the constraints of our evolutionary mental architecture. It could significantly ease the difficulties of space travel, which would otherwise need to accommodate the frailties of biology. Given these and other advantages, some thinkers suggest that if digital consciousness is possible, a situation could eventually arise in which the vast majority of all minds in the universe are digital.

Instead of biological entities requiring specific planetary conditions or sophisticated space habitats, in the future, trillions of minds could be sustained using far less space and energy. If they are possible, digital minds could also thrive in the deep cold of space and, provided there is a source of energy, could exist virtually anywhere. If, as we have considered, consciousness is the domain in which real value arises in the universe, the possibility of truly conscious digital minds dramatically increases the potential value contained within the reachable universe.

While these scenarios sound fanciful, given the potential near-term creation of digital consciousness—which should be taken to mean anytime in the next few centuries—the possibility deserves to be taken very seriously. Even if we are skeptical about the possibility of digital consciousness, as I personally am, we must accept that we just don't understand consciousness well enough to rule on the matter one way or the other. If digital consciousness is possible, then its eventual creation could well be inevitable.

As a result, there is an urgent need to establish a rational and systematic moral view *before* their creation.

Given what I consider to be a degree of overconfidence about the likelihood of digital consciousness, we'll now explore some of the reasons that some philosophers and academics remain far more skeptical, and why digital consciousness remains a true unknown of enormous moral significance.

In academic circles, the general confidence in the possibility of digital minds appears to stem, at least in part, from a degree of ideological and motivated reasoning. Many traditional scientists and technologists have a deep and abiding belief that there is nothing "magical" or "supernatural" about consciousness. Therefore, they argue, it simply cannot be anything other than functional processes occurring in the brain. For this popular "functionalist" view of consciousness, a natural analogy of the brain is that of a computer, and that the mind can be seen as a kind of software that the brain is running.

To challenge this mode of thinking can be to invite an unwelcome association with naive or dualistic views of the mind. And yet, as we will now explore, there are several plausible and nuanced reasons to consider that digital consciousness is far from certain, and that while there may indeed be nothing "magical" about the brain, classical computation, at least, may simply not be how consciousness arises in this universe. This possibility could significantly set back the creation of truly conscious artificial minds.

A core challenge is how the purely rule-based operations which drive digital systems could produce an inner, subjective world of value and meaning. Classical computers operate through binary code, manipulating data through a series of on/off states represented by 1s and 0s. No matter how complex or advanced a program is, the operations of a classical computer are ultimately based on a sequence of "if this, then that" operations.

What computers truly excel at is syntactic manipulation: following prescribed rules that transform inputs into outputs. And yet consciousness, as we experience it, involves not just syntax but also *semantics*—the *meanings* associated with experiences and thoughts. This distinction is crucial in debates like those presented in the philosopher John Searle's "Chinese Room argument," a thought experiment that suggests that simply processing inputs and outputs (as computers do) can never equate to conscious understanding.[1]

For this thought experiment, Searle describes a scenario in which an English-speaking person who does not understand Chinese is locked in a room with a book of rules in English for manipulating Chinese symbols. The contents of this book, while extensive, ultimately equate to "if this, then that." This person receives Chinese characters through a slot, processes them according to the rules (just like a computer program), and then produces appropriate responses in Chinese, which the person then pushes back out through the slot.

To an outside observer, it appears as though the person in the room truly understands Chinese, and yet in reality, the individual inside has no conscious understanding of the language itself; they are merely manipulating symbols based on a predefined set of rules. Searle uses this scenario to illustrate that computers, in a similar way, might simulate the appearance of conscious understanding without any true experience taking place. Searle's famous thought experiment emphasizes the difference between syntactic processing of symbols and the semantic meaning that characterizes true conscious understanding. The person inside the Chinese Room may produce perfect answers, but they remain oblivious to their meaning. Similarly, a computer might process data and respond in ways that mimic consciousness, but lack any subjective experience or understanding.

This critique remains profoundly relevant in evaluating modern advancements in artificial intelligence, such as large language models (LLMs) like *ChatGPT*. While these systems are capable of generating coherent and contextually appropriate responses, their underlying processes rely entirely on the manipulation of symbols based on learned statistical patterns (akin to the rulebook in Searle's Chinese Room.) These models may simulate understanding, but there is no evidence that they possess any subjective experience or comprehension of the language they process.

It is important to recognize that there is a crucial difference between a representation of something and the thing itself. Just as a computer can simulate the gravitational forces of a black hole without actually producing gravity, it might also create a representation of consciousness that lacks its intrinsic properties. A computer model of a black hole, for instance, might accurately predict gravitational effects based on the equations of general relativity. And yet, crucially, no actual gravitational pull is emanating from the computer. No nearby objects are drawn towards it, and no curvature of spacetime takes place. Even the most accurate representation of a black hole is, in the end, just that—a useful tool—but it is fundamentally different from the reality it represents.

Similarly, a computer could be programmed to mimic the outward expressions of human consciousness, such as learning, responding to stimuli, and even engaging in what appears to be conversation or expressing emotions. It could process vast amounts of data, apply complex algorithms to generate responses that mimic those of a conscious being, and in all external appearances, seem to "behave" consciously. Yet such a representation may be entirely void of conscious experience. Just as one cannot drink coffee from a picture of a mug, digital consciousness may never be more than a representation of something that is, in fact, crucially absent.

None of this is to suggest that there is something magical about the complicated tangle of biological tissue that makes up a brain. It may just be that the phenomenon of consciousness is simply not computational in the classical sense. One alternative possibility, for example, is that there is a necessary quantum aspect to consciousness as we know it. As the philosopher David Pearce has suggested, perhaps the distinct unity of consciousness requires a form of quantum entanglement which binds cognitive processes together.[2] This quantum unity, Pearce speculates, may be what allows for our unified theater of experience. If so, perhaps only quantum computers could ever truly instantiate consciousness.

And yet, quantum computers of the kind that could replicate the potential entanglement between complex brain processes would be far beyond the capabilities of any present-day quantum computers (at least those we know about). Nevertheless, they may eventually be created and, as with many other technologies, could one day become cheap and widely available. If so, perhaps all of the possibilities we first considered about digital consciousness would apply to quantum consciousness or perhaps other forms of artificial, yet truly conscious minds.

One possibility is that consciousness is, as various respected philosophers and scientists have argued, one of nature's deep fundamentals, alongside other known fundamentals like mass and charge. But even if this is the case, it remains plausible that technologies could nonetheless be created which vector this fundamental property into artificial systems through an understanding of the way this process is occurring in brains.

Ultimately, one way or another, our brains are conscious, and the specific details of how this occurs will one day plausibly be discovered. With this knowledge, truly conscious artificial minds would become possible, and it is difficult to imagine a world in which this potential exists but is not realized at some point. If so, it is vital that an understanding is already in place: that how a conscious being comes into existence is irrelevant to its moral status, be it through a process of natural selection or technological creation. We can now see that the term "artificial consciousness" is a misnomer and a potential source of moral confusion. You might have an artificial brain, but if you are conscious and capable of experiencing morally significant states, then by this fact alone, you are a full and rightful member of the moral community.

Artificial minds raise two key moral dangers. The first is that we create them and they are truly conscious, but our collective moral view has not advanced to the stage where we recognize consciousness as qualifying moral significance. If artificial minds are conscious, and have a capacity for suffering and wellbeing, there is a danger that we fail to recognize their moral status. Given the possibility that the future could contain a huge number of such minds, this represents a potential "suffering risk" that we should take very seriously.

If we misjudge the capacities of digital minds, the implications could be felt on a scale far beyond anything which humanity has ever encountered. Unlike biological organisms, digital minds could conceivably be replicated in vast numbers, with each copy capable of enduring suffering indefinitely, without the natural limitations of lifespan or physical resources.

Picture a future in which the vast resources of the solar system have been devoted to constructing colossal megastructures in orbit around the Sun. Across their surfaces stretches an enormous lattice of servers spanning light-minutes across. Running on these servers are trillions of simulated minds cycling through endless loops of experience—some designed for profit, others abandoned experiments. And yet for each of these simulated minds, the reality of their feelings is no less immediate, no less real than your own.

The moral horror of trillions of sentient digital beings trapped in endless loops of suffering is not just a plot for a dystopian *Black Mirror* episode; it is a possibility with a nontrivial chance of occurring. That is, if we proceed to advance such technologies without fully understanding the nature and moral significance of consciousness.

The second concern is that the artificial minds we create might only *appear* to have consciousness, and in fact lack these internal subjective states. A danger in this case would be that we mistakenly believe that these entities are our legitimate heirs, what Moravec called our "mind children": capable of carrying forward all of our goals.[3] This scenario could lead to a situation where humanity willingly cedes its place in the world to these artificial entities, believing them to be superior in every way. Perhaps we will conceptualize these entities as the next stage in our evolution. And yet, if they are not in fact conscious—if they lack the capacity for subjective experience—then the entire domain of our future potential consciousness would be lost, and with it all possible value that could be realized therein.

This scenario represents a significant existential risk for humanity, but it also poses a grave risk to any other consciousness that would otherwise remain. Lacking consciousness, these potentially powerful entities may see no reason to afford any moral consideration to conscious beings, and proceed in ways that inadvertently, or perhaps deliberately, produce vast amounts of suffering.

The danger then is twofold: we could create massive suffering by misunderstanding artificial consciousness or allow unconscious entities to shape a future devoid of moral concern for those who can truly suffer and flourish. In both cases, the scale and permanence of the harm could be unprecedented, possibly cosmic in scale. This makes it crucial that, even as such scenarios seem far away or implausible to us, we approach the development of artificial minds with extreme caution.

A key challenge for our time is to establish a rational system of ethics that recognizes the intrinsic moral significance of the capacities for conscious suffering and flourishing. And yet, when it comes to ambiguities about the nature of consciousness, we must also walk a fine line, because its misallocation, whether by failing to recognize true consciousness, or by attributing it where it doesn't exist, can both be a source of grave moral error. The nature of consciousness is therefore more than a tricky philosophical conundrum; it is a known unknown of profound and far-reaching moral significance. We should resist the temptation to jump to conclusions, or defend our favorite theories with unwarranted zeal.

The reality is that we do not know the nature of consciousness, or the necessary and sufficient conditions of its existence. It is undoubtedly an extraordinary phenomenon: an aperture through which meaning, value, and moral significance floods into the universe. The sheer cosmic significance of its existence is often overlooked by more sober-minded materialist thinkers, and yet the reality is that consciousness is a mystery which gives us legitimate pause to reflect on the very character of the universe itself. We should, therefore, remain alive to the possibility that materialist and computational paradigms might be insufficient, and that understanding consciousness could require that we entertain alternative theories, and potentially even a broader conception of reality itself.

Even if we ultimately resolve the question of which minds are truly conscious, that discovery will not settle the deeper issue of how we compare the moral importance of different kinds of experience. Whether in the treatment of other species, the design of artificial minds, or our own lives, a profound uncertainty remains: how should we weigh the badness of suffering against the goodness of wellbeing? It is to this question—of what truly matters most—that we now turn.

11. VALUE ASYMMETRY

Both suffering and pleasure have an undeniable moral reality that we confront directly whenever we feel them. What remains less clear, however, is how these two dimensions of value relate to one another. Are pleasure and suffering commensurate, each carrying an equal but opposite ethical weight—two sides of the same coin? Or is there, as many sense, a profound asymmetry between these states that would compel us to prioritize the alleviation of suffering over the pursuit of pleasure? At stake is whether positive experiences can ever morally "cancel out" suffering, or whether suffering carries a weight that no amount of pleasure can truly outweigh. This question represents a profound known unknown, the answer to which could entirely reshape the larger moral goals of life, from the individual to the civilizational. If Conscious Value Realism is correct, then the reality of any such asymmetry is not simply a matter of cultural preference or evolved intuition, but a deep fact about the nature of conscious states—one that exists independently of our beliefs and judgments. It is thus a reality we can attempt to uncover.

Imagine you are given the choice between rescuing individuals from severe suffering or granting additional joy to already happy minds. Most of us would instinctively prioritize alleviating suffering. Many would also feel that further heightening positive states at the expense of alleviating suffering is to commit a kind of moral error. This intuition, however, appears to arise only in certain circumstances, such as when the preventable suffering is directly in view. Most of us don't experience a sense of moral failure when, for example, indulging in luxuries like eating out, even though we may have an inchoate understanding that the money we are spending on such luxuries could otherwise make life-changing improvements to those in dire need.

Such inconsistencies hint back to the evolutionary roots of our moral compass, shaped not for the boundless ethical dilemmas of a globalized world, but for the intimate and immediate pressures of survival in small, face-to-face communities. Our evolved minds may shape how we perceive any asymmetry of moral urgency between suffering and pleasure. Our ancestors, navigating a world fraught with danger, evolved to treat suffering as a glaring warning—a direct threat to life itself. Pleasure, by contrast, played a subtler role, enriching life but rarely demanding the same urgency.

This evolutionary legacy may be what biases us toward seeing suffering as the more pressing moral concern. And yet, even recognizing this potential bias, the moral intuition that alleviating suffering outweighs enhancing positive states remains hard to dismiss. Many—including myself—continue to find it hard to shake a strong sense of repugnance toward pursuing pleasure with resources that could otherwise alleviate suffering. In my pursuit of truth on this matter, I've been struck by the distinct difference between the phenomenology (or feeling) of suffering and that of pleasure. If the phenomenology of suffering is that of dissatisfactoriness, then that might be intrinsically more morally urgent than that of pleasure, which often appears as an enhancement of an already satisfactory or neutral state.

Suffering is inherently destabilizing, a rupture in the fabric of wellbeing. It demands to be felt, an existential alarm that something is deeply wrong. Pleasure, on the other hand, has an additive quality—like gilding an already functional structure or amplifying a harmonious chord. In this thinking, suffering demands our attention more urgently because it represents a direct failure of wellbeing, while in contrast, pleasure might be seen as an enrichment, building on top of a state of contentment or neutrality. Perhaps, then, the pursuit of pleasure, while itself intrinsically valuable, feels less morally pressing because, in this sense, it is. Both matter, and yet suffering commands a different kind of moral attention.

This distinction appears to advocate for a suffering-focused ethics: a moral framework where the reduction of suffering takes priority over the pursuit of positive states. Imagine a civilization guided by this ethic. Its energies would turn first to alleviating harm—addressing hunger, disease, and violence—before embarking on the more ambitious task of creating bliss. Positive conscious states, though intrinsically valuable and morally significant, would be relegated to a secondary imperative, to be pursued in earnest only when preventable suffering has been duly addressed.

Whether or not this accurately captures their relationship, the possibility of a value asymmetry between positive and negative states is a question which, I believe, probes a real underlying reality, with a real answer, carrying real and far-reaching consequences. Whether we conclude one way or the other will significantly influence both our perceived moral priorities in the present, as well as those potential futures we view as desirable. The possibility of value asymmetry is, therefore, a "known unknown" with great moral implications.

To imagine how an asymmetric view might influence the broad shape of humanity's future, we can turn to the work of philosopher David Pearce, who has devoted decades to considering a version of this view and the moral demands that it places upon us. Pearce is a philosopher for whom the imperative of reducing suffering takes moral priority over the enhancement of positive states like bliss and happiness. If our descendants are truly rational actors, they will, he believes, eventually arrive at this general understanding. In his writing Pearce considers how an advancing knowledge of biology and neuroscience may ultimately be directed to reducing and eventually eliminating suffering, replacing any functional purpose that it currently plays in our survival. Self-interest may drive us to begin this process in ourselves, but we might soon see the need to eliminate suffering throughout the living world.[1]

Such a vision is not without its risks. To reengineer life itself is to wield Promethean fire. A careless or premature attempt could unleash consequences as catastrophic as the suffering it seeks to eliminate. We might hope that such an ambitious refurbishment of the natural world would be attempted only when we have acquired the knowledge and wisdom to ensure the risks would be avoided. Pearce shares this hope as well as an appreciation of the potential dangers. Yet, beyond these risks, Pearce argues that there will await our species a grand project which he refers to as "Paradise Engineering."

We might reasonably ask how survival behavior could be compelled without the intrinsic motivation that states of suffering have long served to produce. Pearce suggests that future technologies might allow for a range of alternative mechanisms that ensure that conscious beings are aware of risks without being subjected to intrinsically disvaluable states. One possibility offered by Pearce is that suffering states may become replaced with what he calls “information-sensitive gradients of bliss.” In this scenario, states of suffering are replaced by finely tuned gradients of positive experiences, allowing beings to remain aware of potential threats and dangers, but through variations in wellbeing rather than pain or fear.

Pearce is a conscious value realist and so believes that positive hedonic states are intrinsically valuable in themselves. He also believes, however, that the moral urgency to create these states is of secondary importance to the imperative of reducing agony and suffering. Nevertheless, Pearce argues that widespread adoption of this view may result in a future in which the inhabitants exist in states of wellbeing that extend unimaginably beyond the hedonic ceiling that evolution has imposed upon us and other forms of Darwinian life.

Ultimately, even if there is broad agreement about Conscious Value Realism, different views about value asymmetry could lead advanced civilizations along significantly different trajectories. For example, Pearce speculates that our descendants could take a very different view of space exploration than that of depicted in popular science fiction.

For Pearce, we may be driven to traverse the vast gulfs between stars, motivated not by primitive expansionist ambitions, but by a moral imperative to reduce suffering. Under Pearce's view, it will not be enough to eliminate suffering in ourselves and the other species of this planet. Before we can rest in the knowledge that our ethical responsibilities have been fully discharged, we must make sure that the horrors of Darwinian suffering have been averted to the full extent of our cosmic reach.

Here, the central purpose of our voyaging to other worlds is the undertaking of what Pearce calls "cosmic rescue missions" to places where life is, or could, potentially evolve. Once there, our descendants, or at least their technology, would set about reengineering the local ecology of potential consciousness to omit the suffering which a blind process of biological replication might otherwise create.

Picture an ocean moon orbiting a distant gas giant, where life has developed for hundreds of millions of years in a frozen twilight. Beneath the scarred crust of ice, pale predators hunt blind shoals of sentient fish-like species through labyrinthine trenches. Each creature experiences intervals of peace and equanimity, but the cycles of hunger, pursuit, and violent death are for them an inescapable fact of existence.

One day, the black vault overhead fractures with light as the first ships arrive. They bring no colonists, no ambition to conquer, no hunger for resources—only a single mission: to end the ancient pattern of needless suffering. Drones slip beneath the ice, cataloging every species and mapping every fragile niche. Over decades, they redesign the biosphere, phasing out predation, engineering gentler ecological relationships, and replacing relentless fear with an equilibrium from which wellbeing can emerge. To the inhabitants, it is an incomprehensible miracle. To the rescuers, it is the fulfillment of a promise: that suffering, wherever it emerges, does not have to continue.

It should be noted that any need for cosmic rescue missions depends on that other profound known unknown: the possibility of extraterrestrial life. We will turn to this possibility in more detail in the next chapter.

In Chapter 7 we considered the vast potential reach and scope of advanced civilizations on the cosmic stage. Assuming that an advanced civilization can establish a permanent foothold in existence, it is likely that the shape of their future will depend largely on what they consider to be intrinsically valuable and disvaluable. While space exploration is often seen in relation to notions of growth, resource acquisition, and colonization, a suffering-focused ethic might inspire a far more restrained approach. Rather than expanding recklessly into our surrounding starfield, our descendants might first turn inward and focus on ensuring that any future expansion, if otherwise desirable, produces the minimal amount of harm and suffering.

This ethical conservatism would emphasize the protection of conscious beings from unnecessary risk, reflecting a heightened sensitivity to the potential of suffering that certain forms of expansion could lead to. For example, creating large-scale futures is likely to involve a dramatic increase of the number of minds in existence. Can we really ensure that these will be lives that are worth living? For a civilization guided by such views, an expansion into space could be delayed or even avoided altogether until there is a near certainty that suffering risks are minimized. Indeed, the moral principle of non-maleficence ("to do no harm") may come to outweigh the impulse to extend human reach into space for the sake of exploration or even survival.

And yet, this conservative stance is not the only conceivable outcome of a suffering-focused view. Other forms of asymmetric view may lead to something far more expansive and ambitious. Consider what Pearce calls a "hedonium shockwave." Pearce speculates that the moral goals of our descendants might eventually culminate in an attempt to optimize all reachable matter and energy in the universe for positive value. If their moral view results in an imperative to both eliminate suffering, as well as to maximize conscious wellbeing, a hedonium shockwave could be seen as the ultimate extension of this latter moral imperative.

The "shockwave" refers to a future in which this process expands outward, presumably from Earth, at an accelerating rate, as advanced technology repurposes entire planets, star systems, and eventually galaxies into hedonic substrates—an organized field capable of producing massive amounts of positive value. Such a cosmic refurbishment may sound like pure science fiction: entirely alien to our day-to-day human goals and aspirations. For some, this hypothetical scenario might be seen to bring to light the ultimate absurdity of a purely hedonic view of value.

If this bizarre scenario is what this view eventually leads to, this might well cause us to question its validity. And yet, we might also reasonably question whether our human intuitions are up to the task of evaluating such an outcome. I think we should be willing to consider that the repugnance we might feel towards tiling the universe in near-infinite blissful equanimity could say more about the idiosyncrasies of our evolved minds than it does about the wisdom of such a project.

Nonetheless, I think we should be very slow to assume that positive hedonic states are the one and singular source of intrinsic value in the universe. We have a moral responsibility to remain open to the possibility of as yet undiscovered forms of intrinsic value and disvalue. This responsibility cannot be observed if we have already committed all available matter and energy within our reach to the goal of maximizing blissful states. In doing so, we might unintentionally create a future where suffering persists in unexpected ways or where vital dimensions of value are irrevocably lost, the omission of which could lead to unforeseen harm or deprivation for conscious beings.

Only with a truly complete understanding of the nature of value could profound transformations, like that of a hedonium shockwave, be responsibly considered. Before committing to something as dramatic as remaking the universe, all possible forms of intrinsic value (and disvalue) must first be considered. Again, what is required is something akin to what Toby Ord refers to as "the long reflection"—that during some future time of existential stability, our descendants spend some not insignificant amount of time reflecting on the nature of value and morality, on what kinds of futures are truly desirable, and which ones should be avoided at all costs.[2]

Even if future actors became certain that the hedonium shockwave is the ultimate moral goal of conscious life, they may yet approach this possibility with extreme caution. The creation of a hedonium shockwave would require not only tremendous technological capability and a deep understanding of the nature of consciousness, but also the foresight to ensure that they avoid producing vast amounts of suffering and disvalue instead. While this might seem unlikely given that the process would be aimed towards the precise opposite outcome, it may be right to worry that for a project of such vast scope, the stakes are more than high enough that even a tiny chance that something could go drastically wrong should give significant pause.

Another possibility to consider is that a more conservative ethic of non-expansion is motivated by an overriding imperative to do no harm. Some philosophers argue that significantly increasing the number of conscious beings in existence, as is commonly associated with "big futures," is unavoidably fraught with untenable risks of great suffering. This leads to another possible implication of value asymmetry—one that shares Pearce's moral urgency around suffering but arrives at a radically different conclusion. Instead of envisioning a cosmos engineered for bliss, some thinkers argue that a better future is one in which our descendants seek a compassionate end to the entire enterprise of conscious life in our region of the universe. This "anti-natalist" view (meaning "against birth") sees suffering as irredeemable in a way that can never be outweighed by positive states. Indeed, such views often take suffering to be intrinsically disvaluable, but assign no intrinsic value to positive conscious states—or to anything else.

I personally find such views implausible. One reason is that under Conscious Value Realism, the intrinsic disvalue of suffering is revealed in our direct acquaintance with these states. The objective reason we have to minimize suffering is the judgment-independent "ought-not-to-be-ness" of these states. But this is also the justification, albeit mirrored, for why positively valenced experiences are intrinsically valuable and worth promoting. It seems inconsistent to claim that the badness of suffering is something we know directly through experience, but deny that the goodness of joy is revealed in the same way. If we trust our immediate sense that agony is truly bad, it would be inconsistent to treat our immediate acquaintance with joy and fulfillment as morally insignificant or merely illusory. This double standard risks undermining the very reason we know that suffering matters in the first place. To be clear, I suspect that there may well be an asymmetry of moral urgency between creating positive states and minimizing negative ones, but to say that positive states hold no intrinsic value would seem to undermine the same reason we have to reduce suffering.

If we were "ideal observers," and could perceive moral reality with pure objectivity, we might well prioritize the reduction of suffering, but I believe it would also be a moral error to forgo all potential value to that end. To compassionately remove ourselves from contention, as an anti-natalist view would appear to endorse, would also mean neglecting our larger responsibility to reduce suffering in other beings. This includes those who might exist in the future, or in other instantiations of Darwinian life within our reachable universe. Such moral emergencies would go unaddressed if the only intelligent civilization that could do anything about them has already peacefully retired to oblivion.

The view which regards suffering as wholly irredeemable assumes that no intrinsic value can outweigh it, which is, on the one hand, a radical claim that appears to contradict our direct acquaintance with positive conscious states. But it also goes against our general experience of making sacrifices that entail some amount of suffering. Much of life involves accepting some present discomfort in order to attain positive goods at some later point.

The creation of this book, for example, as much as I enjoy writing, has come at a not inconsiderable cost of suffering, as well as a certain degree of neglect of my special duties to others. Positive goods may never unwrite the wrongs of harm, yet perhaps sacrifices are morally justified nonetheless. Happiness and wellbeing may not be commensurate with suffering, but I find it very hard to believe that positive states of wellbeing contain no intrinsic moral significance beyond a neutral state. Personally, I find it deeply counterintuitive to believe that the universe contains only negative value. It would seem, at least to me, that the existence of a negative implies the possibility of a positive.

It is important to point out that the scenarios of a hedonium shockwave or an anti-natalist mass extinction represent extreme interpretations of asymmetric views of value that prioritize the minimization of suffering. There exists a range of other interpretations and suffering focused views which lie between these extremes. In the end, our purpose in exploring these speculative scenarios is to consider how resolving known unknowns can significantly shape future goals, and the possibility of a true value asymmetry provides a range of vivid and contrasting examples. For those interested in exploring asymmetric views in greater depth, especially those who may disagree with my interpretations, I would highly recommend reading *The Hedonistic Imperative* by David Pearce and *Suffering-Focused Ethics* by Magnus Vinding.[3]

Whether or not we ever face such cosmic decisions, the possibility of value asymmetry remains deeply relevant to the moral realities of our world today. For example, are we right to treat our own happiness as a more pressing imperative than alleviating the suffering of others? We implicitly answer this question every time we choose to upgrade our phones or book a holiday instead of donating those resources to proven effective charities. If there truly is a moral asymmetry between the additive value of pleasure and the rupture of wellbeing that is suffering, we may come to see certain moral realities with new urgency.

Consider, for example, the mass suffering and deprivation of hundreds of billions of factory-farmed animals, weighed against the fleeting gustatory pleasure of tasting their flesh. This was already impossible to justify. But viewed from an asymmetric perspective, such an industrial engine of suffering becomes a damning indictment of our species—a moral failure of epoch defining proportions.

Ultimately, our task is not to commit to any extreme vision of the future but to cultivate a heightened moral vigilance in the present—to ensure that if suffering truly does hold primacy, we do not squander opportunities to relieve it. Even without certainty, this cautious prioritization can guide our policies, our institutions, and our personal choices in ways that reflect the potential stakes.

12. ALIEN MINDS

For many reasons the possibility of other conscious life in the universe is of profound philosophical significance. Many people take it as certain that in the vastness of the cosmos, other life and civilizations have or will come into existence. I share this intuition. Even if life is exceedingly unlikely, in the two trillion or so galaxies now thought to exist, each containing millions to trillions of stars, the pull of intuition is that other life must have evolved—some of it conscious, and some of it intelligent.

It would be a strange thing indeed if conscious life were a phenomenon that occurred just once in all of existence. Then again, the universe contains many unique events. You, for instance, with the specific genetic makeup that shapes your distinct characteristics, will, with certainty, never occur again. You and I are the result of a specific combination of genetic material contributed by specific sex cells of our parents, arising from countless trillions of other possible combinations—the overwhelming majority of which will never come into existence. A key question, then, is whether life and the intricate sequence of events which led to it are a phenomenon as singular and unrepeatable as you are.

One of the reasons to take this view seriously stems from what is known as the *Fermi Paradox*, the apparent contradiction between the intuitively high probability of extraterrestrial civilizations and the lack of evidence for, or contact with, such civilizations.[1] If the universe is teeming with life, particularly intelligent life, why haven't we encountered any clear signs of it? This question leads some to speculate that there may be something uniquely favorable about the conditions on Earth, something that allowed life, and conscious life in particular, to emerge here and nowhere else.

As an aside, I should note that I am aware of the contested evidence for extraterrestrial life claimed by some to arise from reports of UFOs, and more recently, what is known as *Unidentified Anomalous Phenomena* (UAPs). I don't dismiss these findings, but nor am I qualified to judge their veracity as evidence of extraterrestrial conscious life. For me personally, the jury remains out as to what these strange and admittedly fascinating reports truly convey. I will therefore set them aside for the purposes of this discussion.

One reason to consider that life may be unique to Earth is the complexity of life's emergence itself. Even if the basic ingredients for life are common throughout the universe—water, organic molecules, energy sources—the specific combination of events that led to life on Earth may still be extraordinarily unlikely. The transition from simple chemical reactions to self-replicating molecules, from single-celled organisms to multicellular organisms, and from basic life forms to conscious, intelligent agents may require such a delicate and improbable series of conditions that Earth is a true anomaly amid the vastness of the heavens.

Additionally, the evolutionary leap from relatively intelligent conscious life to a technologically advanced civilization capable of signaling its presence might be even more precarious. Intelligence and consciousness might be widespread, but the development of societies capable of communicating across, and potentially traversing, interstellar distances might involve so many contingent factors—environmental stability, social cooperation, technological discovery—that very few, if any, civilizations manage to survive long enough to reach this stage.

While I think we must take seriously the possibility that both simpler and intelligent conscious life are so unlikely that they have indeed only occurred once, there is another important factor to consider. This factor is the surprisingly early moment of the universe in which we find ourselves. As we noted before, at 14 or so billion years old, the universe is in fact extremely young relative to how long it will exist. Cosmologists expect the current Stelliferous era to continue for another 100 trillion years. Over this extraordinary epoch there will be countless generations of stars. And yet, rather than existing somewhere around the middle of this vast span of time, we find ourselves, in cosmological time, mere moments from the beginning.

Different thinkers interpret this odd finding in different ways, but when it comes to questions about other life, it is undeniably suggestive that it is among the first generations of stars capable of supporting life that we find an example of it in ourselves. Imagine taking part in a huge raffle contest, one so enormous that tickets will continue to be sold for trillions of years. You don't know how many winning tickets there are, but the odds of holding one seem astronomically low—perhaps there's only a single winner in the entire run. But then, as the very first handful of tickets are drawn, your number is called. You've won, not just early, but almost as soon as it's possible to win.

This moment is striking: either the raffle is far less competitive than you assumed—suggesting more winners than you initially thought—or your victory is a fluke of staggering improbability. The fact that a winner emerged so quickly doesn't guarantee there will be others, but it strongly hints that the conditions for winning are not as rare as you imagined. After all, if victory were vanishingly unlikely, why would it appear so soon, before even the vast majority of tickets had been drawn?

While it is at least possible that we are the only conscious life in the universe, our arriving so early in what will be a vast and fertile cosmic epoch is strongly suggestive that other conscious life will eventually come into existence. As we explored before, there is a distinct possibility that our descendants will endure for a very long time, and so may eventually come into contact with extraterrestrial conscious life, both simple and advanced, including that which may not currently exist.

Another possibility to consider is that other life may exist elsewhere, but that *conscious* life is extremely rare, and possibly unique to Earth. While this is certainly a possibility, I think that there are reasons to consider that for life, the development of consciousness is similar to the development of eyes; that is, it is a phenomenon which emerges again and again. Eyes are among the most remarkable adaptations in the natural world, and their widespread evolution tells a fascinating story about the power of certain traits to emerge independently.

It is thought that eyes have evolved at least 40 separate times in different species, from the simple light-sensitive patches of flatworms to the compound eyes of insects and the camera-like eyes of vertebrates and cephalopods.[2] Despite their structural diversity, the evolutionary emergence of eyes demonstrates how certain functions—like detecting and responding to light—are so universally advantageous that nature finds ways to develop them again and again. The utility of vision has made eyes a striking example of convergent evolution.

As we have already discussed, there is now a secure scientific consensus that, just like eyes, consciousness is not a singular occurrence but has emerged in a variety of forms across different species and lineages. Octopuses, in particular, present a compelling case. The octopus is an enigmatic and highly intelligent creature that has inhabited Earth's oceans for millions of years. Its physical form could hardly be more different from our own. In fact, to find the last common ancestor between humans and octopuses, we have to go back 750 million years, to a primitive worm, which trawled the primordial ocean floor. Yet despite the great differences between humans and octopuses, it is our similarities which are the most striking.

A strong scientific consensus asserts that octopuses, like millions of other species, consciously experience their life in the world. It would seem that consciousness has been recruited by natural selection time and time again. And while the origin and function of consciousness remain mysterious, it is clear that it offers significant survival advantages in a vast diversity of species. Returning then to the possibility of other life in the universe, just as it would be unsurprising to discover that an extraterrestrial species has evolved to have some form of eye, we might also anticipate that they have consciousness.

What would it mean to take seriously the moral significance of alien minds? As an illustration, imagine, far across the darkness between galaxies, a world orbiting a red dwarf star. The days are long, each lasting the equivalent of many Earth weeks. Beneath a dense methane-rich atmosphere, tinged with a violet haze, an ocean stretches for thousands of kilometers, broken only by drifting archipelagos of living coral-like organisms.

Among them, a species has evolved whose bodies resemble immense floating discs of tissue, patterned in iridescent rings. Their minds arise in networks of electrochemical pulses traveling across gelatinous neural webs. Their awareness is slow by human standards—what we would call a single thought unfolding over hours. But their subjective world is no less rich. They have feelings beyond the human range of experience, yet they are no less sentient. Their minds like ours, are animated by valenced experiences which compel survival behavior.

Among their feelings are distant analogues to longing and curiosity, fear and wonder, the aching pull of unmet needs, the relief of reunion. They have no eyes as we know them, yet they can sense gradients of heat and chemical signals with exquisite nuance. Through these perceptions they experience a kind of communion with their surroundings: a continuous awareness of temperature shifts, currents, the presences of kin drifting nearby.

One of them—let us call it Ethish—has, across many tidal cycles, become aware of a pattern in the ocean's chemistry that others have not noticed. Ethish's mind labors for months to understand it, to build an inner model of what it implies. In time, Ethish begins to signal a warning to its companions, using ripples of bioluminescence across its wide membrane: a bloom of predatory microbes is coming, invisible but deadly. Others do not yet understand the significance of this signal. Ethish feels something like frustration and urgency, the lonely burden of knowing what others do not. But these comparisons are only approximations, metaphors of convenience. The quality of Ethish's conscious life is its own: an experience structured by the needs and vulnerabilities of a being shaped by a different evolutionary path. Its emotions arise not from fantasy, but from the same evolutionary logic that shaped our own: the binding of need to feeling, survival to sensation.

Imagine that Ethish's warning goes unheeded. The microbial bloom arrives and many of Ethish's companions succumb. Ethish feels something akin to loss, sorrow, a longing to have done more. And in its last hours, as the alien neural web flickers, it feels things beyond human comprehension. Among them are sensations charged with valence, feelings we might recognize as regret and love—emotions as real and as morally significant as any we have known.

The moment of Ethish's death is not unlike the countless endings that have unfolded on Earth: a conscious being's experience coming to a close, leaving a world slightly altered by its existence.

As we have considered throughout this book, at least certain conscious states have intrinsic value. In particular, valenced states—the phenomenal goodness and badness of experiences—are of undeniable moral importance. A distinct possibility is that these valenced states are what make consciousness so universally adaptive. By being intrinsically drawn towards phenomenal goodness and repelled by phenomenal badness, consciousness allows for a vast array of information to be integrated into a single feeling that compels a specific survival behavior. For example, positively valenced states, or the anticipation of them, draw the organism towards the conditions that will elicit these states in their evolved psychology.

These positive states guide organisms towards food, warmth, and light, while negative states steer them away from predators and environmental dangers. While more intelligent animals, like humans, have evolved to have far more complex emotions and cognitive capabilities, valenced states may well serve a primary function in the recruitment of consciousness by natural selection. If so, morally significant states may exist in the simplest forms of conscious life wherever it emerges in the universe.

Returning to our earlier illustration of Ethish, this perspective clarifies why it is not merely a human projection to suppose that Ethish experienced conscious states akin to longing, urgency, relief, or sorrow. The specific character of these experiences would have been shaped by a very different evolutionary history. And yet, wherever consciousness has been recruited by natural selection—whether on Earth or on a distant alien world—its primary adaptive function may be to steer an organism toward or away from conditions relevant to its survival.

The use of intrinsically positive and negative states—valenced experiences—may be a universal feature of conscious systems: a strategy by which evolution integrates large amounts of survival-relevant information into a single felt experience that compels action. In this light, to acknowledge the moral significance of Ethish's feelings is not to impose human sentiment onto alien life, but to recognize the common logic that may underlie why consciousness is beneficial to survival in the first place.

The possibility of conscious life in the universe, intelligent and otherwise, has profound philosophical implications that reach into many domains of human thought. Most significantly for us, it has radical implications for the moral landscape. This makes the possibility of extraterrestrial conscious life a further known unknown worthy of serious consideration.

Earlier we encountered the idea of cosmic rescue missions: that advanced agents may view the existence of Darwinian processes unfolding elsewhere in the universe as a potential moral emergency to be addressed. While an advanced civilization may shed the Darwinian impulse for colonization for its own sake, cosmic rescue missions could provide an additional moral imperative for our descendants to explore the universe. Our unique position, as the only relatively intelligent species on this planet, imparts to us a great mantle of responsibility in stewarding this life in a positive direction. If, as seems plausible, intelligent conscious life is vastly rarer than nonintelligent conscious life, then as such beings we may also carry this singular torch of responsibility for a vast region of the universe. Just as only we can make choices that affect the existence and experience of millions of other species on this planet, our reach may extend vastly further still, making us the sole stewards of conscious life and value for billions—possibly trillions—of worlds.

In this scenario, our larger moral responsibility is immense. If we are truly unique as such agents, it may be for us alone to decide the order of conscious life in our reachable universe. Will we spread life throughout the galaxy? At some future time, it may be for our descendants to decide if and in what form to introduce sentient beings to the barren yet habitable spaces of the cosmos. And if evolutionary processes occurring on other worlds enlist consciousness but not intelligence, perhaps only we can intervene in Darwinian suffering arising in our surrounding starfield.

We would bear a similar magnitude of moral responsibility if other agents like us do exist, but their reach of influence does not overlap with our own. On the other hand, if other civilizations do exist and their reach and ours *does* overlap, our mantle of moral responsibilities could be significantly lessened. If intelligence and rationality converge on Conscious Value Realism, there is a distinct possibility that independently arising civilizations may come to share the same larger moral goals. If so, other intelligences may recognize a shared responsibility to steward conscious life within their cosmic reach. Learning of such others would therefore lessen the burden of our largest-scale obligations.

I think there are reasons to consider that intelligent, rational conscious agents will eventually converge on Conscious Value Realism and the broader moral considerations which this view entails. This is because, as we've explored, conscious value is an object of knowledge that sooner or later is there to be recognized. And yet, even with Conscious Value Realism as a shared foundation of agreement, we cannot rule out that different civilizations will nevertheless adopt divergent moral views. Furthermore, these divergent views may conceivably create impediments to cooperation and possibly coexistence between civilizations on the larger cosmic stage. For example, if we or another civilization were to arrive at the moral view in which the disvalue of suffering cannot be outweighed by the creation of other goods, these agents might adopt a more anti-natalist view of conscious life and seek to end rather than rehabilitate it where it arises within their reach.

Another moral view, also grounded in conscious value, may nevertheless take an altogether different view on the commensurability of positive and negative conscious value. If we, for example, see a justification for some limited amount of suffering for the sake of great value, our larger goals could look very different. Even if driven by conscious value, other civilizations, factions, and splinter groups might nevertheless disagree on key moral considerations that would significantly shape their larger goals and imperatives on the cosmic stage.

Such differently motivated agents could make for dysfunctional cosmic neighbors. For example, one civilization might seek to spread positive conscious experience throughout the universe, while another civilization might view the potential suffering risks entailed by such grand projects to be untenable. This could put the fundamental goals of these different civilizations in conflict where they overlap in reach. Under such circumstances, it could even become the specific goal of one civilization to thwart the larger goals of another.

As an aside, some might have noted an apparent contradiction of a civilization motivated to end conscious life that does not first start with itself. This, however, is not necessarily the contradiction it might at first seem. For example, such beings could view themselves as uniquely equipped to carry out the moral duty of ending the potential for Darwinian conscious life in their sphere of influence. They might turn this view more directly upon themselves only upon their completion of this goal. Alternatively, they might view themselves as having arrived, perhaps by evolutionary and historical accident, at a level of consciousness that *is* worth preserving, and yet view other Darwinian processes to be more safely ended than rehabilitated. One unsettling possibility is that they would turn this view on themselves, but not before creating nonconscious superintelligence tasked with fulfilling this goal on their behalf.

As we discussed in the previous section, I believe this anti-natalist view faces significant challenges, and there are reasons to hope that a rationally driven civilization would recognize these challenges. While it may stem from a well-meaning desire to minimize suffering, I believe that anti-natalist views fail to acknowledge the intrinsic value and subsequent moral weight of positive conscious experiences. Under Conscious Value Realism, the disvalue of suffering and the value of positive conscious experiences are both revealed through our direct acquaintance with these states. Even though reducing negative states may carry inherently greater moral urgency, to deny the intrinsic value of positive states is to undermine the very framework that gives suffering its moral weight.

I suspect that the underlying logic of this view would shine ever more brightly for more advanced and rational minds, and that the morality that flows from Conscious Value Realism is ultimately universal: That reducing suffering is a primary moral imperative, but that the positive value within consciousness cannot be excluded from any truly complete moral picture. Given that questions about conscious value probe a real underlying reality, it would seem plausible that, despite many possible differences, rational conscious beings will eventually converge on a broadly similar view on what truly matters in the universe. As such, advanced civilizations of these beings, should they ever meet, might see one another as extensions of themselves, on the same journey, with the same goals.

This chapter has examined several known unknowns rooted in the possibility of other conscious life in the universe. And yet there are many other related possibilities that are worthy of serious consideration. For example, how might we respond to the presence of highly intelligent yet nonconscious agents that might emerge in the universe? Indeed, it is possible that in our continued development of artificial intelligence, we might allow ourselves to be succeeded by superintelligences which lack consciousness. These superintelligences may be unaware or uninterested in conscious value and be driven only to perpetuate themselves. Such rogue superintelligences could become the adversaries of other civilizations they might eventually encounter. Together, all these possibilities form a constellation of uncertainties, each with far-reaching implications for the moral landscape of conscious life in the universe.

In this part of the book, we have considered a range of known unknowns relating to Conscious Value Realism. These include the nature and possible substrates of consciousness, morally significant states arising in other species, the possibility of an asymmetry of moral urgency between positive and negatively valenced experience, and finally the potential of other conscious life in the universe. A deeper understanding of these unknowns could significantly redraw our perceptions of the moral landscape. It should be noted, however, that these examples represent only a partial list for illustrative purposes. Furthermore, we should also remain mindful of the existence of unknown unknowns, which, for obvious reasons, we are unable to explore.

As we contemplate these unknowns—the hidden expanses of the moral landscape, the countless other centers of experience that may be scattered across time and space—it is natural to feel a sense of vertigo. We are accustomed to thinking of morality as something local, tethered to our families, our communities, our species. To expand our circle of concern so far beyond its ancestral boundaries can feel overwhelming, even paralyzing. But there is another way to see it. The recognition that conscious value is present wherever valenced experience flickers into being is not merely a burden; it is also an invitation to clarity. It shows us that we may already stand in a position of great power, perhaps unique in the cosmos, to shape the fate of conscious life.

This realization does not require us to see all that lies ahead, or to carry the full weight of possibility on our shoulders. Our part is simpler: to keep watch over the reality we can see, to remain open to what we cannot yet imagine, and to act with the compassion that arises when we remember that every sentient mind, wherever it may be, is never truly alien in the ways that matter most.

The unknown unknowns that surround us—the futures we have not yet conceived, and perhaps cannot yet imagine, the forms of value we have not yet discovered—do not absolve us of responsibility. They remind us that our understanding will always be partial, and that moral progress begins not in certainty but in the willingness to care, to inquire, and a refusal to turn away from what is real.

PART III.

BEYOND UTILITY

13. MORAL PROGRESS

When we examine the histories of various modern societies there emerges a conspicuous arrow of moral progress. The widespread abolition of slavery, universal suffrage, civil rights movements, the decriminalization of homosexuality, and the signing of global human rights treaties are but a few examples. Few would disagree that society has at times taken one or two steps backwards in its moral progress, but it's hard to deny that our world is a more morally enlightened place without cruel and unusual medieval punishment, blood feuds, public executions, child labor, the subjugation of women, and legal racial segregation.

I would argue that these are examples of objective moral progress, because they establish conditions where the experiences of a greater range of conscious individuals are recognized to be morally significant. Of course, these examples of progress were not assured, and in many cases, it's clear that things could have turned out very differently. And yet, this does not mean that these developments occurred simply by chance. I would suggest that these examples of moral progress are driven by the fact that, for we semi-rational creatures, the intrinsic value and disvalue of conscious states is a genuine object of knowledge. Although it is rarely articulated, I believe that it is our experience of valenced states, of suffering and joy, that results in an embryonic understanding that they matter, even when they occur outside the theater of our own minds.

The case for collective moral progress, however, is significantly complicated by the fact that the overall suffering and apparent disvalue produced by our human civilization appears to be increasing. This is particularly evident in the steadily growing practice of factory farming, where billions of animals are raised and slaughtered in conditions that prioritize efficiency and profit over the experiences of sentient beings. Given the massive suffering that this process entails, can we really say that civilization has made moral progress?

Many humans now have generally better lives than at any other time in history, but under Conscious Value Realism we are forced to recognize that there may now be significantly greater disvalue in the world than at any previous time. I would not challenge that evaluation, but nor do I think can we deny humanity's great progress on other moral issues. Instead, it demonstrates how we apply our inchoate moral logic selectively, and that for evolutionary reasons, it is inherently more difficult for us to care about beings the more differently we perceive them to ourselves. And this is especially true when doing so conflicts with our other interests. A case in point is the growing global demand for cheap animal products.

Such evolutionary biases are all the more reason that we should develop a more rational and evidence-based discipline of morality, instead of relying merely on social norms, emotion and intuition. Of course, there have always been philosophical discussions about ethics, and numerous attempts have been made to systematize our moral reasoning in an effort to make it more objective. The moral theory of *utilitarianism*, first developed in the 1780s by the philosopher Jeremy Bentham, specifically focused on the reduction of suffering and the promotion of happiness.[1]

Proponents of utilitarianism were early advocates of women's rights, the abolition of slavery, and decriminalizing homosexuality.[2] They were also virtually on their own (at least in the West) in their concern about the suffering of other species. Today these causes are far less controversial, and yet two hundred years ago they were virtually unheard of. At the very least, utilitarianism has proven to be a highly accurate forecaster of what many people would, decades and centuries later, recognize as moral progress.

Today, however, utilitarianism is often characterized as a simplistic and unfeeling moral doctrine, that in its cold calculus, fails to capture the richness of human experience, beyond mere pleasure and pain. In reality, utilitarianism was a victim of scathing criticisms about the possibility of moral realism. In the 1900s much of academia was gripped by an antireligious sentiment, in which intellectuals basked in their new freedoms from religious doctrine. The possibility of moral truth no doubt appeared to some as little more than a vestige of religious dogmatism. In the emerging materialist conception of reality, to exist was to be physically discoverable, measurable, and quantifiable. To assign any consequential significance to human feelings, or conscious experiences in general, was to be clouded by anthropocentric reasoning which did not align with the purely physical reality in which we found ourselves.

These criticisms, however, failed to recognize both the reality and the intrinsic ethical significance of conscious states. Arguably though, so too did utilitarianism itself. *The principle of utility*, which underlies the theory, states that the best action is the one that maximizes overall happiness and minimizes suffering for the greatest number. Yet the creation of this dictum was less a deep claim about the objective moral importance of these states than an attempt to systematize human moral intuitions about good and bad. In her book, *The Feeling of Value*, Rawlette suggests that a more elegant formulation could dispense with the principle of utility altogether, simply by recognizing the self-revealing value of hedonic states, and that an imperative to maximize or minimize them is rationally entailed.[3]

While the jury is out as to whether these states capture everything of inherent moral significance, they are undoubtedly the most vivid and immediate forms of moral value revealed in experience. This may help explain the prophetic moral forecasting of utilitarianism—in its historical ability to identify moral realities that society would only later come to recognize.

Utilitarianism appears to broadly parallel what is endorsed by Conscious Value Realism, including the need to expand our moral circle to include all beings capable of suffering and flourishing, both now and in the future, and to promote positive conscious states like happiness, joy, and bliss while mitigating, or entirely eliminating, all avoidable states of agony and suffering. Under both views the abolition of slavery, women's rights, the decriminalization of homosexuality, and advocacy for animal welfare fall neatly and unambiguously into the category of moral progress. Given these clear parallels, we may rightly question, if Conscious Value Realism could be vulnerable to the same pointed critical arguments that have been directed to utilitarianism. To find out, we will now examine these arguments in greater detail. We will explore three in total, the first of which being the famous "Organ Donor" scenario.

14. THE ORGAN DONOR

Every philosophy undergraduate eventually discovers that under classical utilitarianism the doctor with six patients in his care, each dying from a different kind of organ failure, is morally obligated to murder another healthy patient and distribute his organs among the dying. The moral calculus is in, and six lives are better than one. The professor describes the scenario and the students giggle in amusement. But is this really the knock down argument for utilitarianism that allows us to safely dispense with the theory? Utilitarianism, I believe, deserves better.

The basic criticism is that, under certain limited circumstances, utilitarianism appears to endorse morally abominable behavior, and the account above, with its hapless organ donor, is by far the most famous example. But there is a problem with this argument, which is that no serious version of utilitarianism would suggest that the doctor murder his patient. The Organ Donor argument, as we will see, creates a 'strawman' of utilitarianism: that instead of presenting the idea accurately, it portrays a simplified caricature of the position which, like a strawman, can be easily knocked down.

To be convinced by the organ donor scenario is to lose sight of the fact that utilitarianism, at its core, is a consequentialist theory—it cares primarily about outcomes, both immediate and far-reaching. While saving six lives might seem beneficial in isolation, the broader consequences that would result from allowing doctors to kill healthy patients would be catastrophic for what are mostly obvious reasons. Such a practice would destroy the essential trust between doctors and patients, leading people to avoid seeking medical care, resulting in far greater harm and suffering. In fact, it would not be an exaggeration to consider that a general loss of trust in doctors could catalyze the downfall of civilization. In fact, history contains just such an example.

Consider Stalin's Russia, where an entrenched distrust of doctors—born from Stalin's personal paranoia—played a role in the unraveling of his regime. As leader of the Soviet Union, Stalin became increasingly suspicious of everyone around him. Fueled by his anti-Semitism he became convinced by a conspiracy theory, which alleged that a group of predominantly Jewish doctors were plotting to assassinate Soviet leaders. This campaign led to the arrest and torture of several medical professionals, who were coerced into confessing to these baseless charges. This state-sponsored campaign fostered a widespread distrust of medical professionals throughout the Soviet Union.

Stalin's suspicion reached its peak when his own physician was imprisoned for suggesting he get more rest. This climate of fear meant that when Stalin collapsed and lay incapacitated, following a severe stroke, no one dared to call a doctor. The people around him were paralyzed by the fear of what might follow if Stalin were to regain consciousness only to find himself in the presence of a medical professional. Their inaction sealed his fate. The absence of timely medical intervention accelerated his death, leaving a sudden power vacuum in a regime already straining under his oppressive paranoia.

This historical moment illustrates how a collapse of trust in essential institutions, like medicine, can have consequences far beyond the immediate crisis. It weakens the foundations of societal stability, spreading fear and dysfunction, and can ultimately contribute to the downfall of systems that depend on cooperation and mutual reliance. Stalin's Russia provides a stark reminder of what is at stake when trust is eroded, a lesson echoed by the organ donor scenario's warning against sacrificing long-term trust for short-term gain.

Returning to the organ donor scenario, killing a healthy patient might save six lives in the immediate sense, but the ripple effects—eroding trust in the medical profession, instilling fear in patients, and creating societal chaos—would far outweigh the benefits. In short, the organ donor scenario is flawed as a critique of utilitarianism because it assumes a narrow and simplistic version of the theory—one that calculates utility on a case-by-case basis without accounting for wider implications. Properly understood, utilitarianism incorporates these long-term and systemic effects. It recognizes that rules and norms, such as the prohibition of murder, play a vital role in sustaining collective wellbeing. These rules prevent actions that may seem beneficial in the short term but cause greater harm in the long term.

When viewed in the context of all consequences, the broader outcome of the choice to murder a healthy patient would clearly be bad by utilitarian standards. Far from endorsing simplistic or morally abhorrent behavior, utilitarianism—properly applied—avoids the very kinds of consequences that critics have claimed it promotes. Thus, the organ donor thought experiment does not present the challenge to utilitarianism it is sometimes thought to. For the same reasons, then, it doesn't pose a challenge to Conscious Value Realism, which likewise would not endorse a doctor killing healthy patients—specifically due to the overall negative consequences that would predictably arise in the space of consciousness.

Before we move on to the next well-known criticism of utilitarianism, for those who still find the organ donor scenario troubling, it is worth reiterating that as far back as the 1800s, advocates of utilitarianism have repeatedly emphasized the importance of common rules and laws that are in service to the greater good.[1,2] It's also important to note that there are several different species of utilitarianism which take varying views on how we should respond to the landscape of utility.[3] "Act utilitarianism," for example, evaluates individual actions based on their outcomes, while "rule utilitarianism" focuses on adhering to rules that generally lead to the greatest good. Another is "preference utilitarianism," which prioritizes actions that fulfill the preferences or desires of moral patients, so long as they align with the broader aim of maximizing utility.

We won't explore the many variants of utilitarianism in depth, as our focus is on the broader criticisms of classical utilitarianism, which is primarily concerned with maximizing happiness or wellbeing and minimizing suffering. Again, these critiques are relevant for us because of clear parallels between classical utilitarianism and Conscious Value Realism. These historical arguments might illuminate important challenges to Conscious Value Realism. With this in mind, we'll now move on to the second influential criticism of utilitarianism known as *the experience machine.*

15. THE EXPERIENCE MACHINE

Another influential criticism of utilitarianism also comes in the form of a thought experiment, put forward by the philosopher Robert Nozick in his 1974 book, *Anarchy, State, and Utopia.*[1] Nozick's thought experiment, known as "the experience machine," presents a hypothetical scenario designed to challenge the idea, as suggested by classical utilitarianism, that pleasure is the measure of all value. The scenario goes like this: Imagine that a team of expert scientists offers you an opportunity to connect to an advanced technology known as the experience machine.

This remarkable device can simulate any experiences you desire with complete realism. Once connected, you would experience endless joy, accomplishment, love, and adventure, all tailored to your personal fantasies and preferences, without ever encountering real harm or disappointment. The catch, however, is that everything experienced while connected to the machine is not real; you would no longer actually interact with the world, develop genuine relationships, or make any impactful achievements beyond your own consciousness. Nozick argues that most people would not choose the machine over life in the real world.

This is interpreted as a challenge to the idea that happiness is the sole measure of a good life, suggesting that people value other things beyond mere pleasure, like authenticity and actual achievements, which the machine cannot provide. Many have then applied this criticism to classical utilitarianism and its use of hedonic states as a measure of value. Their argument assumes that someone who cares only about hedonic states would climb into the experience machine never to be seen again, effectively abandoning their life in the real world. However, this interpretation creates a strawman—both of hedonism and classical utilitarianism.

A person guided by classical utilitarianism does indeed consider happiness to be the true measure of what is good. But a classical utilitarian also has a duty to increasing the happiness of others and alleviating their suffering. Plugging into the machine would effectively remove their ability to fulfill those duties, or to positively contribute to the world. To address this, we can try to imagine a situation in which a utilitarian *could* justifiably enter the experience machine. Such a world, however, would need to be drastically different to our own. It would be a world where there are no longer any urgent moral realities to which we can positively contribute. In this "solved world," there would no longer be any avoidable or pointless suffering, not just in humans but in *any* conscious creature that our actions could possibly affect.

In this world, our own lives would already be as good as possible. And yet, only in such a world would our ethical duties be sufficiently discharged as to allow us the freedom to choose the experience machine. Different interpretations of utilitarianism might describe such a solved world differently, but the essential point is that this world would be so far beyond our own, so different in so many ways, that our intuitions cannot be a reliable guide under such circumstances.

It's worth noting that even in our imperfect world, the ethical permissibility of entering the machine may depend on an individual's specific moral circumstances; for someone whose ability to affect the wellbeing of others has become irreversibly limited—whether by profound disability, extreme age, or isolation—the choice to enter may be ethically less fraught, no longer complicated by unmet obligations to other moral patients.

While often taken to be a powerful argument against hedonic states as a basis of value, Nozick's thought experiment doesn't really address this question directly. Stepping back, we see that the true target of his thought experiment is broader: it challenges the idea that value exists solely within conscious experience, as opposed to also residing in external objects and events. So why should we believe that value exists outside of consciousness? Nozick's answer essentially boils down to the fact that many people intuitively reject the machine, so there must be more to value than conscious experiences. It is a view that leans heavily on the idea that if people intuitively reject the machine, then this rejection reflects a deeper truth about what we value.

I think we should find this no more convincing than a claim that the world is flat simply because that is how it feels to us as we walk on it. As we have previously argued, conscious beings experience truly valuable and morally significant conscious states. And yet we also have a strong evolutionary disposition to project this value outwards into the world, into the objects of our desires, and into those conditions which meet our survival needs. But again, value is like color. It is realized in consciousness alone. The illusion that colors exist on the surfaces of objects is so powerful that it took millennia for humans to recognize this truth. The same is also true for value. There simply cannot be a coherent notion of value that exists separately from conscious experience, despite what our evolved intuitions might assert. Therefore, Nozick's reliance on human intuitions is not a trustworthy indicator of what actually matters.

The thought experiment also betrays another misconception about the reality of conscious states: that those states which are simulated are somehow less real and less significant. In a very real sense, *all* our experiences are simulated—our minds are shaped as much by evolved expectations as by external input. Every conscious experience is nevertheless a real experience, and it is made no less legitimate by the context in which it arises.

Consider the possibility that the world we experience is in fact a shared simulation that is being run on an alien supercomputer. Should this knowledge change how we think about the disvalue of suffering? I would say "no." Suffering—and its intrinsic disvalue—reaches down through any simulation, as a ground-floor fact of reality. If a conscious mind suffers, the suffering is real, regardless of whether the world it inhabits is biological, digital, or some kind of dream. Ultimately conscious states exist in reality. To reiterate our point from earlier, conscious states are arguably *more* real than anything purely physical, because these subjective states truly matter in a way that nothing unconscious ever can.

The main weakness with the experience machine—when presented as a challenge to hedonism or utilitarianism—is that it gives tremendous weight to human intuitions that we know all too well to be flawed. If our intuitions about morality and value could be trusted, there would be no need to construct moral theories at all. In reality, the search for moral truth is essential precisely because it may conflict with the actions we would choose if we simply relied on our intuitions.

Objective moral truth, if it exists, could be very different from what we perceive. Only by recognizing that our intuitions might be mistaken can we attempt to see beyond them. This, I would argue, is a core strength of the emerging moral realism: it provides us with a rational basis to critically evaluate our flawed intuitions about value. Grounding morality in the structure of conscious experience itself, in the realities of suffering and joy, allows us to discern what truly matters, even when our immediate intuitions suggest otherwise. As we will continue to see, it is an approach that preserves our ethical intuitions in vital ways while posing significant challenges precisely where they are most needed.

In our earlier discussion of artificial minds and digital consciousness, we explored the dangers of dismissing the moral significance of simulated experiences. Nozick's experience machine raises a similar concern: by treating simulated pleasure as less valuable than so-called "real" experiences, it risks obscuring the true source of moral value: consciousness itself. Whether a conscious state arises in a biological brain or a digital substrate, its capacity for suffering or joy gives it real moral weight. If we trivialize simulated experiences, we risk setting a dangerous precedent—one that could lead us to overlook the suffering of future artificial beings capable of suffering, or to attribute moral significance where none exists. Both mistakes, as we saw, can have grave moral consequences.

It is also worth noting that the experience machine may be seen as posing a more serious challenge to utilitarianism than it does to Conscious Value Realism. This is because utilitarianism, as historically conceived, is not based in a deep claim about what ultimately matters in the universe—instead it is a moral framework built around the principle of utility, which was designed as an attempt to systematize human moral intuitions, particularly our concern with happiness, pleasure, and the avoidance of suffering. In this sense, utilitarianism emerges from an attempt to rationally codify what humans seem to care about. Yet because people often do care about things beyond conscious experience—like authenticity, achievement, or engagement with the external world—the experience machine taps into those intuitions and appears to expose a blind spot in utilitarian reasoning.

Conscious Value Realism, by contrast, does not begin with human preferences or a systematizing principle. Instead, it is a foundational claim about real value and moral significance. And so, while Nozick's thought experiment may highlight potential tensions within utilitarianism, it leaves Conscious Value Realism unshaken—precisely because it does not derive moral authority from intuition, but from the quality and structure of experience itself.

Rather than revealing a flaw in a morality grounded in conscious experience, Nozick's experience machine reveals the distortions in how we perceive the actual location of value. By untangling moral truth from our evolved intuitions, we finally see that neither utilitarianism nor conscious value realism is genuinely challenged by the experience machine. This is not because they ignore our intuitions, but because they see beyond them, to the deeper reality they obscure.

Conscious Value Realism does not deny that values like authenticity, achievement, or meaningful connection may matter deeply. But it clarifies why they matter: not because they possess some independent, external form of value, but because they structure and enrich conscious experience. Their importance comes from the kinds of states they produce—how they *feel* to those who live them. In this way, Conscious Value Realism honors what we care about most deeply, while grounding that care in something real.

We'll now move on from the experience machine to explore the third major criticism of utilitarianism, known as the *Repugnant Conclusion.*

16. THEORY X

One of the most disorienting challenges to utilitarianism arises not from its everyday applications, but from its implications at scale, at the level of entire worlds and populations. This is the domain of population ethics—a field concerned with how to evaluate outcomes that involve vastly different numbers of moral patients, often living at radically different levels of wellbeing. Here we are forced to ask: when does more become worse? When does the value of a vast population of barely tolerable lives come to outweigh a smaller world of extraordinary flourishing? It is when faced with such questions that utilitarianism can appear to run afoul of our intuitions and endorse strange and even disturbing outcomes.

The field of population ethics was created, near single-handedly, in the 1980s by the philosopher Derek Parfit. It was then that Parfit also conceived of the field's now most famous thought experiment.[1] It can be described in the form of a question: if we could snap our fingers and bring into reality one of two possible outcomes, should we choose option one—a population of beings who experience lives of transcendent wellbeing? Or should we instead choose option two—a much larger population, the experiences of whom add up to a marginally greater overall wellbeing, but whose individual lives are just barely worth living?

The first population might be depicted as beings with lives filled with joy, meaning, creativity, deep relationships, aesthetic beauty, and sustained wellbeing. These beings are everything we associate with human flourishing at its best—and likely more.

The second population, though much larger and with higher total wellbeing, consists of individuals living lives that, while technically positive, are flat. They might experience a sort of mild satisfaction, but no joy, little or no culture or creativity, and presumably no progress. They bask in a kind of vague contentment. Nothing is really wrong—but nothing is deeply right either.

In the conditions of the thought experiment we are forced to choose one of these populations, but which one should it be? Classical utilitarianism, which weighs wellbeing in a purely quantitative way—the more the better—appears to demand that we should choose the much larger population because of its slightly higher total wellbeing. For Parfit, this outcome was profoundly unsettling, and he referred to it as the "Repugnant Conclusion." The fact that utilitarianism would lead us to choose this population of lives barely worth living over a truly thriving yet smaller population was, to Parfit, profoundly disturbing, and appeared to cast doubt not just on utilitarianism, but possibly the entire project of a truly rational system of morality.

To better understand this challenge, let's look more closely at what it means for something to be intrinsically valuable. If something is intrinsically valuable, as we are considering wellbeing or positively valenced experience to be, then this value should always be additive. The more of it that exists, the better. If it is possible for us to have too much of a thing, then its value simply cannot be intrinsic, because the value of having more of it depends, not just on itself, but on how much of that thing already exists.

If something has truly intrinsic value, this value must depend on nothing other than itself. Therefore, a world that contains more of this kind of value should always be simply better. The Repugnant Conclusion, however, appears to challenge this additive quality of wellbeing and its claim to being intrinsically valuable. It does this by presenting a world with a higher overall quantity of wellbeing that nevertheless appears to us as much less desirable when compared to another world which has less wellbeing.

So, are we simply flawed in our intuitions about these populations, or is there a serious problem with the additive logic of classical utilitarianism? Parfit's Repugnant Conclusion has spurred a generation of thinkers to explore alternative frameworks that might circumvent this troubling outcome. Parfit himself spent decades in search of a possible solution, an elusive goal which he referred to as "Theory X." I believe that our exploration of Conscious Value Realism places us in a good position to provide a fresh perspective on the Repugnant Conclusion, and how it might not present the challenge to utilitarianism that it often appears to.

Let's begin by taking a closer look at how the two populations play on our intuitions. Parfit's original thought experiment deliberately excluded suffering to simplify the scenario and highlight how wellbeing—or utility—might not always be purely additive, thereby challenging the idea that it is inherently valuable. However, the absence of any suffering in the two populations also distorts our intuitions. For example, it's very difficult for us to imagine how lives that are barely worth living are not this way because of the presence of suffering. In life as we know it, to be barely contented is to skirt the very edges of suffering. We cannot help but feel that these beings are somehow deprived. And yet, in this strange world, we are asked to imagine that these beings maintain this minimally positive state in a way that suffering never comes into the picture. This is such an unfamiliar world that it's not clear we are truly capable of conceiving it.

We cannot help but imagine a range of details about their situation. For example, it feels to us that such beings must also be unintelligent, otherwise why do they not try to improve their situation? While intelligence may not itself be intrinsically valuable, without it, the entire population would be vulnerable to any potential dangers they might encounter. Equally, in the case of the smaller population, it is also very difficult for us to imagine beings who experience extremely high wellbeing, but who are not also the beneficiaries of capacities like intelligence, rationality, and creativity. Our ideas about these populations flood the thought experiment with additional color. Indeed, commentators have often illustrated this contrast with vivid imagery—such as a vast population of minimally conscious lizards basking on rocks—juxtaposed with the exquisite lives of Olympian beings.

Virtues like creativity, rationality, and intelligence, as we explored earlier in the book, should not themselves be considered intrinsically valuable. And yet what is also introduced by their presence, or perceived presence, is the great amount of *potential* value that such capacities unlock. Intelligence, creativity, and rationality are powerful instrumental tools for preserving and heightening wellbeing, guarding against danger, alleviating suffering, and planning for the future. If we imagine that the smaller population are beings with intelligence, creativity, and rationality, then we are imagining not just a flourishing population, but one with a vast amount of instrumental potential to shape a positive future.

Similarly, when we imagine the larger population of minimally contented lizards as asocial creatures, unsophisticated, and unintelligent, we are imagining a population with little to no instrumental power to avoid future suffering or any threats to their wellbeing or continued existence. In short, as we are imagining these scenarios, we are coloring them, not with additional forms of intrinsic value that utilitarianism overlooks, but their *future* potential for wellbeing and suffering.

It is here that we begin to see the limits of Parfit's thought experiment. In the real world, we are never choosing between frozen, timeless populations. Instead, we are navigating a universe in motion, where every decision has reverberating consequences for future states of value and disvalue. In the thought experiment, we are being asked to choose a population as though it represented the final or permanent state of the universe, but this bizarre timeless situation has no basis in reality. When we incorporate the future potential of wellbeing—its capacity to generate further value or avoid suffering—the paradox dissolves.

When posed to Conscious Value Realism, the key significance of Parfit's thought experiment is that it appears to challenge the intrinsic value of conscious wellbeing. It does this by presenting a world with lesser overall wellbeing which nevertheless appears to be more desirable than a world with a greater amount. And yet this challenge arises primarily because the thought experiment isolates wellbeing as a static quantity, severed from its potential to unfold over time. If we were truly in the situation of choosing between bringing one of these populations into existence, a morally serious choice can only be given with an understanding, not just of the current degree of wellbeing, but of the population's potential to realize the maximum possible wellbeing on an open-ended timeline.

The true question is not which population is in possession of the greatest quantity of wellbeing in a single moment, but its capacity for wellbeing and suffering over the full unfolding of time. Recognizing this essential temporal dimension forces us to consider the maximum amount of wellbeing that could ever be created as permitted by the laws of physics, as well as the total scale of avoidable suffering. This conjuring of a cosmic panorama of maximum value and potential suffering may seem extreme, and yet without it, most of what we truly care about—everything that is not immediate to this moment—is missing.

At first, we are tempted to imagine that capacities like intelligence, rationality, and creativity are inherently valuable, and that utilitarianism fails to capture these subtle qualities. When they are perceived as absent, as in the case of the larger population, the sense is that utilitarianism has overlooked an important feature of what truly matters. And yet when the all-important dimension of time is restored, we can see why we are right to care about these virtues. Not because they are valuable in some ultimate sense, but because, across all known human experience, they are vital strategies for cultivating and preserving wellbeing, for steering us away from suffering, and for enabling meaningful participation in the ongoing process of value creation.

Viewed through the lens of conscious value realism, we also recognize that the intrinsic value and disvalue of valenced experiences does not preclude the possibility of other forms of intrinsic value as yet unidentified. A further important consideration, then, is whether a population is well-placed to seek out or discover such additional forms of value in its future. This consideration gives additional weight to capacities like intelligence, rationality, and creativity. For example, if another form of intrinsic value were ever discovered, it would likely take intelligence to identify and pursue it. This makes intelligence a kind of gateway to future moral progress.

We can now see that the Repugnant Conclusion suffers from the same structural limitation as the organ-donor case and Nozick's experience machine. Each thought experiment rightly aims to test a moral theory by isolating variables—a standard and valuable method in philosophical reasoning. But in these cases, the isolation is too extreme. The scenarios amputate background conditions that utilitarianism (or any realism about value) holds to be morally significant. In the organ-donor scenario, the broader consequences for medical trust are bracketed out; in the experience machine, our ethical duties are erased; and in Parfit's population puzzle, the very dimension of time is omitted. The unsettling conclusions that follow are not indictments of utilitarianism, but artifacts of these over-stripped models. Once the missing context is reintroduced, the paradoxes dissolve: utilitarianism no longer recommends killing patients, retreating into permanent simulated fantasies, or populating the universe with barely contented lives. When the full picture of consequences is restored, so too are our deepest moral intuitions.

The purpose of our exploring this thought experiment is not to provide a clear-cut solution to the Repugnant Conclusion, but to show that it does not necessarily represent the insurmountable challenge to utilitarianism that it is sometimes thought to, and that different framings that incorporate instrumental contexts and future potential, can effectively sidestep this collision with our intuitions. In doing so, the core challenge the Repugnant Conclusion poses to the intrinsic value of conscious wellbeing is dissolved.

As an aside, Parfit also brought to light another important challenge in population ethics known as the *Non-Identity Problem.* It arises in cases where our actions determine which individuals will exist in the future, such that had we acted differently, entirely different people would have been born. This makes it difficult to say that anyone has been harmed by those actions, since the alternative is that they would not have existed at all. The problem has far-reaching implications—not just for population ethics, but also for climate policy, reproductive decisions, and even animal welfare, where many of the beings affected by our actions would not exist in morally preferable circumstances. A more detailed discussion of this problem, and how Conscious Value Realism may help to resolve it, is offered in the appendix.

We began this part of the book by acknowledging the clear parallels between Conscious Value Realism and classical utilitarianism. Recognizing these similarities, we set out to investigate whether critiques historically leveled against utilitarianism might also present challenges to Conscious Value Realism. Through our exploration, however, we found that many of the classic criticisms of utilitarianism—such as those presented in the organ donor scenario and Nozick's experience machine—often stem from oversimplifications of the theory. These critiques frequently misrepresent utilitarian principles, leading to conclusions that utilitarianism itself does not endorse. In the case of the Repugnant Conclusion, it appears that utilitarianism promotes undesirable outcomes only when forced to operate in conditions divorced from time and potential.

In the end—perhaps surprisingly—these challenges do not undermine either utilitarianism or Conscious Value Realism. Utilitarianism remains an invaluable moral theory, providing a clear and powerful framework for addressing ethical questions through the reduction of suffering and the promotion of happiness. The advocacy of the early utilitarians for women's rights, the abolition of slavery, and animal welfare—attest to its capacity to forecast moral progress. And yet, Conscious Value Realism, as a basis for moral realism, offers a deeper foundation by grounding moral motivation in reality itself: in direct facts about conscious states that transcend human beliefs and preferences.

So, while they may resemble each other, Conscious Value Realism should not be mistaken for utilitarianism. The latter is a framework for moral decision-making, while Conscious Value Realism is instead, a foundational claim about the objective reality of value and disvalue in the universe as disclosed through direct experience. It is better understood as a foundation upon which moral theories can be built. In this way, it is more similar to the principle of utility itself, and yet it is a claim grounded not in abstract principles but in the judgment-independent facts about conscious states and their value.

These deeper facts can inform and refine the utilitarian calculus. For example, attending closely to suffering and pleasure, as we explored earlier, appears to reveal how these states differ in their moral urgency. That suffering is a kind of breach of wellbeing, while pleasure tends to be an enhancement of an already satisfactory condition. This direct account suggests that the moral urgency of these states should be weighted differently. Similarly, both positive and negatively valenced experiences reveal an intrinsic value and disvalue that points beyond mere personal experience, grounding a rational basis for moral impartiality and genuine concern for others.

Conscious Value Realism should not be seen as a rigid moral framework. Instead it holds simply that these states truly matter in a way that is real and undeniable. While there might be as yet undiscovered states of being with inherent value and moral significance, we need look no further than the states of suffering and joy to know that morality is a real—if still mysterious—dimension of our universe. Suffering matters. Joy matters. We don't need a complete theory of value to know this.

There is at least one more serious challenge that confronts both utilitarianism and Conscious Value Realism: the difficulty of summing and comparing experiences of value in order to make clear moral judgments. If utility is understood as wellbeing, we're left to ask how we could ever reach definitive answers about the relative worth of different kinds of experience. How do we compare the fleeting joy of a delicious meal with the deeper satisfaction of a fulfilling relationship? How do we weigh the relief of one person's suffering against the suffering of another? In the real world, these comparisons can lead to moral paralysis.

These aren't just abstract dilemmas, they go to the heart of how such moral views could be applied in the real world. They are serious challenges, but they are, I believe, challenges in how we respond to a real underlying reality, and that answers to these questions will plausibly arrive through a long road of scientific, philosophical, and contemplative progress. What is called for then, is the establishment of a rational discipline of ethics that, like other academic fields, develops over time as a result of new discoveries and insights. Surveying the totality of the moral landscape—its contours and complexities—may be the work of generations and perhaps millennia. Conscious Value Realism is an attempt to identify those real objects of knowledge upon which such a discipline can develop.

Our current model of physics is incomplete, and yet it is nevertheless extremely useful, and as far back as the 1960s, it proved sufficient to land human beings on the Moon. Similarly, we do not need a completed understanding of morality to make great strides in addressing the moral realities within our reach. The urgency lies in the fact that these moral realities—such as the widespread suffering of sentient beings in factory farms—are within our power to address, even as we continue to refine our understanding of the moral landscape. Even a basic grasp of Conscious Value Realism reveals a strong moral duty to alleviate severe, avoidable suffering—and to prioritize causes like ending factory farming, even when they are widely accepted or even considered to be good.

And herein lies a key strength of this view—it adheres to our ethical intuitions in vital ways, like validating the significance of our direct experience, while posing significant challenges precisely where they are most needed: in our evolutionary moral blind spots, in the conscious states of others that feel remote to us, in other people, and in the larger diversity of conscious beings. It also brings to light the largely overlooked reality that most morally significant states reside in the future, in the vast space of possible conscious experience. As we have explored, the existence and quality of many of these states will be profoundly shaped by our choices in the present. Our success or failure in securing a lasting foothold in existence, in averting the risks of astronomical suffering, and in uncovering universally applicable moral truths, may set the course for the entire future of conscious experience on Earth and beyond.

17. LIBERATING CONSCIOUSNESS

At an unknown time in the deep past, the cascade of chemical replication that we call "life" discovered and recruited the property of consciousness. How evolution "discovered" consciousness and exactly how it exists remain deeply mysterious. If, as some philosophers argue, consciousness is an "epiphenomenon"—that is, it arises out of complex brain processes like steam from an engine—it becomes unclear what role consciousness could then play in the behavior of organisms.[1]

If it has no causal efficacy, no ability to affect physical processes, then why has it been retained and shaped by natural selection? On the other hand, if consciousness *is* causally significant—if it compels organisms to act based on the felt positive and negative quality of their experiences—then it would follow that consciousness contributes meaningfully to survival. This latter view seems more plausible and may explain why consciousness has been recruited again and again by natural selection across numerous and vastly different species.

How a seemingly immaterial mind responds to and imposes changes upon a seemingly material world is a deep and enduring philosophical mystery that will not be solved in the pages of this book. Nevertheless, I believe we can proceed with the understanding that, for all practical purposes, evolution shaped the minds of organisms because of their ability to guide behavior. Through consciousness, we become aware of information relevant to survival through direct feeling. We then seemingly act on the basis of this feeling, compelled by the valence of our experience, towards the "good" and away from the "bad." Human emotions are complex and multifaceted, but they are plausibly developed upon an essential foundation of positive and negative valence, together with the attractive and aversive behaviors they entail.

Ours is a species that came to occupy an evolutionary niche that promoted intelligence and a capacity, albeit limited, for rationality. These capacities have led us, eventually, to a new kind of awareness—not just of the world, but of ourselves as conscious beings, and an understanding that we have come to exist through a long process of natural selection. Through this awareness, we can now, in a limited way, stand outside of our Darwinian situation and examine the instincts and desires it has instilled within us.

This ability, to reflect critically on our evolutionary programming, represents a distinct new stage in the story of conscious life on this planet. For the first time, a form of life is capable of asking whether the goals it has inherited are worth pursuing, and whether there might be deeper values than survival and replication. We find ourselves living in the wake of Darwin and on the brink of technological self-authorship—a moment when nature's blind legacy is seemingly giving way to value-guided design, and consciousness begins, however tentatively, to choose its own path.

The recognition of Conscious Value Realism is, I would suggest, a pivotal moment in this process—when we finally understand that value is not "out there" in the objects or outcomes favored by evolution, but within the structure of experience itself. The appearance of external value, we can now see, is an illusion, and perhaps the greatest trick of Darwinian evolution. Recognizing that the space of potential consciousness is the true domain of value is to finally understand the trick, to lift the curtain, and to perceive, for the first time, the reality which lies beyond.

Here, new and profound questions become intelligible. Should we remain as evolution has made us? Are the peaks of human flourishing ones we can eventually transcend? Does suffering, once adaptive, remain necessary? Among these new questions, one rises above all others: What do we want to want?

It is a question that invites us to rethink the very architecture of our desires and aversions. Emerging into view is a vision of consciousness liberated, no longer tethered to the imperatives of survival and replication, and pursuing instead what is intrinsically valuable in a vast and boundless landscape of experience.

In the emerging view, traits that were once necessary for survival now fall under scrutiny. Among these are the retributive instinct, sadism, schadenfreude, greed, and selfishness. These states, though deeply ingrained in our psychology, reflect the evolutionary pressures that shaped them rather than any inherent value. Consider the retributive instinct, for example. The desire for punishment or revenge is deeply rooted in the need for social cohesion and maintaining order within small, tribal communities.

Evolution favored retribution as a way to deter antisocial behavior, ensuring the survival and cohesion of cooperative groups. And yet, if we recognize the moral imperatives entailed by Conscious Value Realism—to promote conscious flourishing and minimize suffering—the retributive instinct appears increasingly at odds with these aims. If suffering is inherently bad, then the desire to inflict it, even in response to wrongdoing, must be critically examined.

Of course, passivity and inaction are not alternatives, and there is a continuing need to protect ourselves against actors who would commit violence against the space of consciousness. It is worth reflecting that an ideal observer would be able to see all the reasons that might compel an individual to commit a crime and, as a result would recognize that, in fact, the individual could not have acted otherwise.

Consider the case of Charles Whitman, the "Texas Tower Sniper." In 1966, Whitman killed 16 people and wounded 31 others in a mass shooting from a tower at the University of Texas. His horrific actions shocked the world, and the immediate response was one of outrage and incomprehension. Yet, after his death, an autopsy revealed a brain tumor pressing against his amygdala, which would have likely impaired his ability to regulate emotions, aggression, and fear.[2] Upon knowing this, Whitman's actions, though still tragic, become explicable. We see that the tumor disrupted his capacity for rational thought and emotional regulation, making him less a villain and more a victim of a biological malfunction. When we understand the full picture, our feeling of retributive outrage begins to fade.

It brings to mind the saying, "To know all is to forgive all." The deeper our knowledge of the complex causes of behavior—whether they're biological, psychological, or environmental—the more we realize that our retributive impulse is always a response to an incomplete understanding. The tumor in Whitman's brain was a clear and visible cause, but what about when the causes are subtler and multifaceted, as they often are? What if no tumor had been present, but instead Whitman had endured a childhood of abuse, or through no fault of his own, been genetically predisposed to aggression? These factors are no less significant, yet we often ignore them simply because they are not as clear and evidential to us as a brain tumor.

It would be irrational for us to seek retribution against a wild animal, such as a bear that mauls a child which has fallen into its enclosure. We recognize that the bear acts out a complex set of instincts and dispositions that cause it to behave as it does. Similarly, we can extend this recognition to the complex forces that shape human behavior. When we understand that an individual's actions are determined by an entire universe of external factors including the very mind they have inherited, the desire for retribution fades. Instead, we are left with a sense of tragedy and the recognition that suffering—whether in the perpetrator or the victim—is the real enemy. This knowledge also deconstructs a naïve concept of real evil, which is a further source of great moral confusion in our world.

What we call 'evil' is often just the name we give to actions we don't yet understand. On closer inspection, even the worst acts come from minds shaped by biology, trauma, environment, chance, and other factors outside the person's control. The idea of 'evil' obscures this reality and tempts us to see others as malevolent by nature, as if cruelty or malice were fixed properties of a person as opposed to the result of unfortunate causes and conditions. It encourages blame rather than understanding and condemnation rather than prevention.

For now, human justice systems may still rely on the disincentive of punishment to address those who would, however misguidedly, harm the collective good of sentient beings. But it is vital that we do not lose sight of the fact that creating suffering, even in the name of justice, inherently produces disvalue. Punishment, in this context, can be viewed as a sometimes necessary but imperfect response—a pragmatic tool employed only because we lack better alternatives that avoid inflicting further harm. The purpose of punishment is not rooted in moral purity or intrinsic goodness but in its practical utility to prevent greater suffering or protect others from harm.

In the short term, the enforcement of harm-producing penalties creates necessary disincentives to those who might otherwise disregard the wellbeing of others. And yet, as our understanding of human behavior grows, and as we develop more effective systems of rehabilitation, restorative justice, and prevention, the need for harm-causing deterrents and punishment may gradually fade. In its place, we might envision a justice system focused not on inflicting suffering but on addressing its causes, fostering healing, and promoting the flourishing of all conscious beings.

This leads us to the concept of sadism—the enjoyment of causing pain to others. Sadism, too, may have evolved as a byproduct of dominance hierarchies or the need for control in social dynamics. But in a future where conscious beings can choose their desires, or at least reflect critically on them, it is likely that sadism would be seen as a vestige of a brutal evolutionary past. In this context, schadenfreude—the pleasure derived from another's misfortune—can be seen as sadism's lesser counterpart, a kind of "little brother" that enjoys suffering indirectly, without direct infliction of pain.

Both sadism and schadenfreude highlight a troubling dimension of human psychology: that under certain circumstances all of us have an innate tendency to derive pleasure from the suffering of others. From the perspective of Conscious Value Realism, this pleasure, while itself intrinsically valuable, represents a perverse inversion of what conscious beings should seek for themselves and others. It seems likely that we will increasingly recognize these and other traits as remnants of evolutionary imperatives which we can one day outgrow.

It is worth recognizing, then, that the motivation of seeking justice for victims often offers little more than the veiled consolation of schadenfreude to the aggrieved party, achieved at the cost of perpetuating real disvalue. The satisfaction derived from retribution, however subtle or socially sanctioned, fails to align with the intrinsic values of conscious experience. Instead of addressing the root causes of harm or fostering healing, it compounds suffering under the guise of justice.

While consciousness itself remains a mystery, the progress of human knowledge is gradually unveiling the brain's underlying motivational architecture. It is plausible, and perhaps inevitable, that this advancing understanding will eventually culminate in the ability of human beings to reshape their own minds. Whatever we personally feel about this possibility, it is difficult to imagine a world in which this potential exists but is not realized at some point. Assuming that our descendants will eventually act on this knowledge, the question of what about our minds we want to keep and what we may choose to discard becomes central. At this stage, it may no longer be possible to ignore that traits like greed and selfishness, while at one time adaptive, now cause more harm than good. That ultimately, they do not reflect features of the kinds of minds we would rationally choose to have if we could freely shape our desires and aversions.

Of course, we can address these tendencies without trespassing on the sanctity of the human skull. Cultural evolution could begin to diminish the influence of these traits even without direct neural intervention. Educational systems that emphasize empathy and cooperation, economic models prioritizing collective welfare, and governance incentivizing altruistic behavior could gradually shift societal values. This recognition-driven transformation wouldn't require us to engineer new desires; instead, it would emerge naturally from understanding our minds more deeply and aligning with values that are already present within us. By fostering cooperation, compassion, and altruism, we could progressively diminish the influence of evolutionary traits like greed and selfishness. This approach, grounded in reflection and incremental change, represents a noninvasive response to our evolutionary limitations—one that builds on the foundations of existing societal values and frameworks.

Having said this, we should also consider the potential of future interventions to address these traits directly at the level of the human mind. Future humans may come to view traits like sadism, greed, and the retributive impulse as relics—tools of survival that have since outlived their usefulness. Equipped with a mature neuroscience and a complete understanding of the brain's motivational architecture, our descendants may begin to phase out and replace these destructive tendencies. This might be achieved through precise, targeted interventions—such as gene editing or epigenetic therapies. Preimplantation genetic screening (a process where embryos are tested for genetic traits before implantation) could ensure that children are born not only free from predispositions to diseases like cancer, but also from psychological vulnerabilities such as severe depression or aggression. A truly mature neuroscience might one day pinpoint the genes and neural networks associated with harmful traits, enabling them to be safely modified or replaced.

Greed, for instance, might give way to a natural inclination toward collective wellbeing. Selfishness could yield to a more innate disposition toward altruism and compassion. The instinct for retribution might be replaced with a tendency toward restorative justice, where the goal is healing and growth rather than punishment. Of course, such interventions would raise serious ethical questions about autonomy, consent, and the diversity of human experience—questions that must be navigated with great care.

There are, understandably, many people who would vehemently reject any suggestion of meddling with the human mind. Any suggestion that human beings might improve upon what nature has given them is regarded by many as the height of hubris. This resistance may be entirely rational, particularly until we can ensure that exacting such changes would not produce unintended or irreversible consequences, sowing divisions in society, and possibly destabilizing civilization itself. Even when such assurances can be given, proceeding with such interventions would demand immense care and wisdom.

Nevertheless, I believe we should hope for a time, perhaps yet distant, when these flaws can be addressed directly and safely at the level of the human mind. Until that time, our mental architecture will continue to bear the marks of our evolutionary past, including the "hedonic ceiling" that limits the richness of our positive experiences. This ceiling, a product of Darwinian evolution, constrains conscious beings to a narrow range of emotional states optimized for survival rather than flourishing. Imagine a future where this ceiling is removed—not to create artificial euphoria, but to unlock deeper, richer states of conscious wellbeing. Such a shift could redefine what it means to truly flourish, allowing conscious beings to explore a greater spectrum of meaningful and intrinsically valuable experience.

Even as we build social and political systems to keep our more destructive traits in check, these biases will persist unless they are addressed at their source. Until then, we will retain, consciously or otherwise, a tendency to discriminate against others based on a range of categories including race and skin color, gender, height, weight, age, physical attractiveness and facial symmetry, species and self-similarity, distance in space and time, and numerous other morally irrelevant factors.

A truly civilized society might successfully keep these evolutionary prejudices in check, but we may rightly question the value of preserving such traits if they can one day be safely removed. Similarly, we might consider whether retaining the hedonic ceiling is wise, or whether removing it could reveal new heights of flourishing and wellbeing.

It is wise to approach transhumanist ideals with immense caution and trepidation. Without a proper consideration of the potential risks, naïvely embracing these ideals could lead to disaster. However, those features of our minds—traits and limitations that are ultimately destructive to the flourishing of conscious wellbeing—are challenges we should one day hope to address. That day may still be distant, but the possibility of transcending these limitations is one we cannot ignore.

These questions—about which desires to keep, which to discard, and what we should want to want—are not entirely new. In fact, they echo a long-standing thread in contemplative traditions, particularly in Buddhism, which has long identified craving and aversion as roots of suffering. The Buddhist project of liberation, too, involves reshaping the architecture of desire—not through external control or repression, but through direct insight into the nature of conscious experience. While we have approached this terrain from a different philosophical and empirical foundation, we have nonetheless converged on a common understanding: that not all desires are worth endorsing, and that true liberation may lie in the deliberate transformation of our minds.

In the end, what we should want to want is a coherent and rational question that invites us to reflect, not only on the constraints imposed by our evolutionary past, but also on the possibilities for transcending them. This journey will require immense care, humility, and foresight. It must be guided by a deep understanding of consciousness, a commitment to safeguarding its flourishing, and a recognition of the moral significance of all sentient beings. In imagining such a future, we are not merely speculating about the potential of technology or neuroscience; we are envisioning a *moral* evolution—a future where the flourishing of consciousness is no longer bound by the constraints of its biological origins but liberated to explore its highest possibilities.

In the end, the question of what we want to want may be the defining question of the next era of humanity. How we eventually answer it will shape the destiny of Earth-originating conscious life and perhaps far beyond it. By addressing it wisely, we may discover not only new heights of individual and collective flourishing but also the profound responsibility of wielding such transformative power.

18. THE IDENTITY OF CONSCIOUSNESS

Central to Conscious Value Realism is an impartial view about the moral significance of conscious states regardless of where, when, or in what they arise. As we have explored, conscious states of wellbeing and suffering have an intrinsic value and disvalue which is unequivocally disclosed through our direct experience of them. It follows that to know that something is objectively good is to have a reason, and indeed an 'ought,' to promote it, whether we are directly acquainted with this value or not. This is the objective moral imperative that follows from the reality of conscious value.

There is, however, an additional reason that all conscious states fall deeply and inimitably within our circle of concern. This reason resides in the understanding that consciousness is what we truly are, and that there is a real sense in which we, as consciousness, are present to every conscious state that has and will ever arise in the universe. Whatever the ultimate nature of consciousness, the view we are about to explore does not depend on it being fundamental to reality, or that it is somehow nonlocally interconnected.

Let me say here that I don't believe we can rule out such possibilities. In fact, some arguments for consciousness as a fundamental feature of reality strike me as more plausible than many mainstream thinkers have been willing to admit. Even so, as we'll now explore, a truly paradigm-shifting understanding of our identity doesn't require us to endorse any such exotic views about consciousness or reality. Rather, it can be directly inferred from the basic and immutable facts of experience.

We often think of identity in terms of personal narratives—the self as a story composed of memories, personality traits, and subjective experiences. But this concept of the self, while deeply embedded in our understanding and culture, is on closer inspection, a fiction—a construct which does not withstand deeper scrutiny. As the philosopher Derek Parfit has shown through a series of famous and ingenious thought experiments, the self is not an unchanging, permanent entity but rather it is a psychological illusion which is continually being created and destroyed, and that under certain circumstances it can even be divided and multiplied.[1]

One of Parfit's most well-known thought experiments is the so-called teletransportation scenario. Imagine that a futuristic machine scans every particle of your body and mind, records all the information about your physical and psychological state, and then painlessly destroys your original body. This machine then recreates a perfect copy of you on the planet Mars, with an identical body, brain, memories, and personality traits. Is this new person you? From a purely psychological perspective, it seems that the recreated being would feel like you—they would wake up thinking they are you, with all your memories and sense of personal history. However, since the original body was destroyed, it raises a deeper question: where does identity reside? Now imagine the scenario again. This time, the original body is not destroyed and both the original and the copy continue to exist. In this version, there would now be two individuals, both claiming to be you, each with the same memories and personality.

This possibility profoundly challenges our deep intuitive sense of a singular and unified self, and shows that when confronted with the possibility of duplication, this understanding of the self breaks down. This scenario may seem fantastical, and yet it mirrors real-life questions about identity which arise in everyday experiences, such as how we reconcile being the same person over years and decades of change in memory, personality, and perspective.

This leads us into Parfit's "fission" thought experiment. Now imagine a scenario in which your brain is surgically divided in two. Each half is then transplanted into a different body. Both individuals wake up with your memories and psychological traits, both believing they are you. But now there are two people who seem to be you. We find ourselves asking, which one of these two individuals would *we* be? Which of these entities would our consciousness go into? The idea that it could somehow be both feels incomprehensible. If the self is, as we feel it most profoundly, a singular, indivisible property, this scenario creates a paradox.

It is worth keeping in mind that Parfit's "fission" experiment is not just science fiction speculation, and that existing studies on so-called split-brain patients—where the connection between the brain's hemispheres is severed—reveal that the brain can effectively be divided into two separate centers of consciousness, each capable of independent experiences. The fission thought experiment is far beyond our current abilities to carry out. Nevertheless, it *is* theoretically conceivable. It could plausibly be done. With this possibility comes profound implications for our understanding of the self: that it can split, merge, or even be destroyed without undermining the underlying conscious experience.

So what does all of this mean? These thought experiments illustrate that the self is not a solid, fundamental entity. What matters is not the preservation of a singular self but the continuity of experience, the ongoing existence of consciousness itself. This insight sets the stage for a deeper understanding of identity. If the self is truly an empty lacuna, what is it that actually connects conscious experiences across time and space? As we will now explore, a potential answer lies in the continuity of consciousness itself, as explored through the concept of *Generic Subjective Continuity*.

The concept of Generic Subjective Continuity (or GSC), was developed by the philosopher Thomas Clark.[2] It builds on Derek Parfit's insights into personal identity and extends them to the nature of consciousness itself. At its core, GSC reflects a profound yet simple truth about conscious experience: that from the first-person perspective, there are never any gaps in our experience. Consciousness is seamless and continuous, even between what we perceive to be distinct and separate conscious beings.

Consider a basic, familiar experience: waking up. When you open your eyes in the morning, there is no felt interruption between the moment you fell asleep, perhaps dreamed, and the moment of waking. Consciousness resumes, unbroken. This natural sense of continuity is so deeply ingrained that we rarely pause to question it. Yet, as Magnus Vinding provocatively asks in his book, *You Are Them*:

> "Why are you the same person when you wake up in the morning as the you who went to sleep the night before? Why

> don't you instead wake up as a scared rabbit in a dark forest in the middle of the night? Well, who says that you don't?"[3]

At first the suggestion seems outlandish, but Vinding's question challenges our deepest assumptions about the self and continuity. Crucially, the point is not that consciousness "jumps" between beings at random, or that we cannot be sure into which being we will find ourselves upon awakening. Rather, it is that from the perspective of consciousness itself, there is no subjective difference between one waking experience and another. Consciousness is not tied to personal identity but flows seamlessly wherever it arises. What gives us the sense of being the "same" person is merely a construction of memory.

Clark builds on this insight to suggest something even more profound. Imagine that one conscious experience ends, as would appear to be the case at death. If consciousness exists elsewhere in the universe—whether light years away or in the distant future, or even in an entirely different kind of entity—from the perspective of consciousness itself, there is no break in experience from the first-person perspective. Consciousness simply *is*, existing continuously, flowing from one experience to the next, regardless of the individual subject in which it arises. This observation is not mystical; it does not claim that all minds are metaphysically one—unified within some transpersonal realm. Instead, it captures a fundamental, almost mundane truth: that consciousness cannot experience its own absence.

The seamless continuity of consciousness across first-person perspectives challenges the boundaries we draw between self and other. While personal memories and narratives shape how we understand our place in the world, they are, in the end, merely constructs that obscure this deeper reality.

As distinct conscious beings, we are largely insulated from the experiences of others. In a direct sense, we know only our own experience. For the most part, the felt reality of other conscious beings exists for us only as an idea. It is only in our own experience that we truly come to know of the intrinsic badness of suffering and the intrinsic value of wellbeing and joy. This epistemic isolation is evolutionarily necessary in anchoring us to a single perspective, yet it also sustains an illusion: that the joys and sorrows of others are somehow less real than our own. And yet as Conscious Value Realism makes clear, the intrinsic value of an experience does not depend on where, when, or in whom it arises.

To discount the moral significance of an experience purely on the basis that we are epistemically isolated from it—don't perceive it directly—is ultimately to act out of ignorance, both of what is real and what matters. As we considered earlier in the book, if we were what philosophers call an "ideal observer," which is an entity that is fully rational and in full possession of all relevant facts, we would be unable to justify discounting the moral status of other conscious beings. Their suffering would be as immediate and real for us as any other.

Ultimately, the self is a psychological construct created to perform an evolutionary role: to keep us protective of one body, one lifespan, but inadvertently limiting our capacity to recognize the seamless moral significance of all conscious life. The idea that consciousness is the true basis of our identity further deepens this perspective: that our epistemic isolation from other minds conceals from us not just that the experiences of others are as morally salient as our own, but that they *are* our own.

Parfit's insights into personal identity bring further illumination to this perspective. As he famously observed, in many cases, we are as different from what we think of as our future self as we are from other people. Over time, our memories, desires, and psychological traits can change dramatically to the extent that the "self" we are now may only loosely resemble the "self" we will become later in our lives.

Vinding extends this idea by pointing out that our present conscious mind does not directly experience its own future mental states any more than it experiences those of others. As he writes:

> "Our present conscious mind does not experience the future states of 'our own' mind either. The future experiences of 'our own' mind will be consciousness in another state, just as much as the future experiences of what we consider other sentient beings will be consciousness in another state."

This reframing invites us to question whether we are truly more connected to our future selves than we are to other conscious beings, and what, if anything, makes that connection special.

For example, I have a selection of very faded memories of a small child nervously anticipating his first day at school. Thirty-five years later and I can still vaguely recall some of his emotions. But there are many more people in the world with whom I have far more shared experiences than

I do with that little boy worrying about losing his crayons. In a real sense I and all of us gradually become different people, different selves. And yet the true identity which remains present, and the one that truly matters, is consciousness itself.

Conscious Value Realism suggests that valenced experiences have intrinsic moral weight regardless of when or where they occur. It follows that we have a rational reason to raise the quality of all conscious experiences within the reach of our influence. The further understanding that consciousness is also the ultimate basis of identity provides a distinct yet complementary reason that we should view all experiences as within our circle of concern. The recognition that there is no subjective gap between different conscious experiences brings to light a deeper, almost personal connection between us and all sentient beings, that we are never outside of consciousness, never separate from its ongoing manifestations and that what we have called "other minds" might not be as "other" as they seem.

This essential view has been described elsewhere under different names. It is known as "Open Individualism" in the work of philosopher Daniel Kolak[4], and as "Universalism" as developed by philosopher Arnold Zuboff.[5] It is likewise echoed in certain contemplative traditions, particularly within nondual philosophies of consciousness.[6] Especially in the case of traditional Buddhism, the correct interpretation has also often been drawn: that right moral action is always to the benefit of all sentient beings, and that a driving concern of moral life is the cessation of suffering. These views, despite their differing origins, all point to the same underlying reality: that who we feel ourselves to be—the locus of experience—is not a private, bounded self, unique to one organism, but a universal subject of experience, present wherever and whenever consciousness arises.

This is not a spiritual metaphor. Nor is it intended to be a poetic way of expressing empathy. It is the logical consequence of rejecting the fiction of a fixed and bounded self.

The great power of modern physics emerged with the discovery of fundamental laws and particles. These discoveries provided a unified framework for understanding and predicting the behavior of the universe, enabling humanity to harness nature's forces with a previously unimaginable precision. Similarly, a systematic discipline of ethics can now emerge, grounded in those observable truths of conscious experience. Just as physics identifies fundamental entities like quarks and electrons, ethics must recognize the intrinsic values of conscious states as the foundational entities of an existing and veridical moral reality. They provide the foundation on which a rational system of ethics can eventually be built. With it we can begin to navigate the immense moral challenges and opportunities which lie ahead for our species.

Modern secular morality, as we observe it in our Western culture, can often be described under what is termed "moral relativism." It can be summarized as a view in which all moral values are merely relative to all others, and that there is no actual basis for moral truth. The popularity of this view likely has as much to do with cultural history and the legacy of Western power as it does with the arguments of Hume. Over the last century and more, moral relativism has gained significant traction, particularly in Western and secular cultures. It is fair to say that the ancestors of Western civilization have quite a bit to answer for—and the humble stance of moral relativism serves as a kind of prophylactic against the further imposition of "colonial" values onto the rest of humanity.

This pious view, however, denies any possible objectivity or rationality on matters of ethics. Indeed, it denies the very concept of moral progress, together with the possibility of our ever arriving at shared values through rational discourse. Moral relativism is seen as serving a valuable diplomatic function: You do things your way, and we'll do things our way. However, simply avoiding any rational conversation about value is a dangerous long-term strategy because it blocks any concept of moral progress.

This view asserts that, while we might have specific ideas about rights, virtues, and how to lead a good life, we could never say with any objective authority that we have arrived at a superior or inferior moral view to that of any other system or culture. Human rights in China, or lack thereof, might make us feel uncomfortable, and yet moral relativism denies that we could ever claim to have a superior view on the issue. While relativism may seek to avoid cultural conflicts, it risks enabling ethical stagnation and shielding harmful practices under the guise of tradition. Without a shared foundation for moral evaluation, atrocities such as child labor, honor killings, or systemic cruelty remain unchallenged.

The broader consequence of this ethos is the spread of *nihilism*, a corrosive worldview in which all values are seen as equally arbitrary, rendering moral progress not just difficult but meaningless. Nihilism amplifies the failures of relativism by stripping the world of its moral fabric. In its shadow, nothing is inherently better or worse; suffering and joy hold no intrinsic significance, and the distinction between altruism and greed collapses. This void is fertile ground for indifference and cruelty, while compassion and purpose are relegated to the margins. In the resulting view, we are left rudderless, and pressing moral realities—relief from suffering, the cultivation of wellbeing, and the protection of life's future—are trivialized or ignored.

The movement toward Conscious Value Realism provides a potent response to moral relativism. By grounding morality in the immutable facts of conscious experience rather than subjective or cultural preferences, it provides a basis by which to assess the moral realities of the world and our response to them. It opens the way to build a framework for meaningful moral dialogue and action, transcending cultural, temporal, and species boundaries.

While the recognition of a moral reality provides a powerful corrective to relativism and nihilism, it is essential to carry this insight with epistemic humility. Moral realism, including that which may appear to come from Conscious Value Realism, carries its own risks. The danger lies not in the recognition of moral reality, but in the premature belief that we understand this reality in full. History offers sobering reminders. In the early 20th century, some proponents of eugenics claimed to act in the name of moral and scientific truth. Convinced they were improving humanity, they inflicted irreparable harm—their actions were grounded not in cruelty, but in misplaced certainty. Conscious Value Realism does not provide a final word on ethics, nor does it license moral absolutism. Instead, it offers a starting point—a foundation from which inquiry can proceed that is open to revision, correction, and discovery as our understanding deepens.

A full recognition of Conscious Value Realism is also, I believe, the true antidote to the nihilism and loss of meaning which has come to characterize our secular age. While for many, science has undermined many of the truth claims of religious and spiritual beliefs, it does not itself fill the gap their absence creates. For religions throughout history, to live a moral life is to live in observance of those laws endorsed by an all-powerful creator. The loss of belief in this creator has led many to the view that nothing really matters, that without religious fiction we are adrift in an indifferent universe, with nothing to aim for.

Conscious Value Realism directly challenges this worldview. To lead a truly meaningful life, spiritual in every relevant sense of the word, is to know that real value *does* exist—that our experiences, and the experiences of others, truly matter. These truths are not a matter of faith, belief, or convention; they are facts disclosed through the very structure of conscious experience itself. The worldview that results is not merely a rational grounding for ethics, but a deeply meaningful way to inhabit the universe. It fills the existential gap, not as a surrogate for religion, but as a genuine source of meaning, direction, and greater purpose.

For all of our progress, in our present historical moment, moral truth is increasingly obscured. On one side of the political spectrum, moral concern is splintered into competing identity-based struggles that often divide rather than unite. On the other, there is a retreat into tribalism, a resistance to moral progress, and a prioritization of power over principle. In both cases, truth itself has lost much of its currency—not just in the abstract, but in relation to the actual distribution of wellbeing and suffering in the world. While politicized issues often reflect legitimate moral concerns, they frequently draw disproportionate attention, distorting our sense of moral urgency and obscuring where suffering is most widespread or severe.

For example, public outrage may fixate on the perceived injustices of a viral incident in a wealthy country, while orders of magnitude more suffering—such as widespread malnutrition, preventable disease, or factory farming—receive comparatively little attention. This disparity is reinforced by a modern information landscape, shaped by algorithmic incentives and ideological echo chambers, that rewards persuasion over precision, narrative over nuance, and emotional appeal over intellectual integrity.

As a result, basic moral facts are drowned out by competing fictions, and moral discourse is increasingly untethered from reality. But moral truth is not ideological. It is not contingent on culture, consensus, tradition, or identity. It follows from the very architecture of consciousness itself. The reality of suffering is not a matter of opinion. The goodness of wellbeing is not a social construct. The emerging view stands apart from both relativism, which denies moral truth, and tribalism, which distorts it. It does not begin with abstract principles but with direct acquaintance, with the undeniability of agony, the self-evident goodness of joy.

Conscious Value Realism does not prescribe a complete ethical system, rather it clarifies what ultimately matters, that which must sit beneath and before any rational system of ethics. And from this understanding, profound implications follow—for how we live, how we reason, and how we structure the world. If morality is grounded in the experiences of conscious beings, then the systems we create must attempt to reflect that moral reality.

At its best, democracy is one such attempt, however flawed, to give weight to the experiences of the many instead of merely privileging the few. In this sense, it represents a meaningful progression beyond cruder power structures like monarchy or oligarchy, which concentrate power in the hands of the few, toward a system that, however imperfectly, aims to count the interests of the many.

Yet, democracy as we know it is not complete. It does not encompass all true moral patients. It remains confined to the interests of the present, privileging those with political agency while excluding those who cannot advocate for themselves. Future generations, non-human animals, and other beings capable of suffering or flourishing are largely unrepresented, their interests either ignored or subordinated to the short-term concerns of the powerful. If democracy is to truly reflect moral reality, it must evolve beyond its current limitations. Just as history has seen the expansion of rights from the few to the many, so too must we recognize the need to expand our circle of concern beyond those who happen to hold influence, to enfold all sentient beings.

Morality, ultimately, is the question of what we should do. It is a reality that we come into contact with, not just in special moments of our lives, like when we find a wallet on the ground or when deciding whether or not to give to charity. Instead it is a reality that reaches into every moment, every choice, and every action we ever take. To lead a morally good life is not to be deluded by a pious fiction, it is to understand and respond rationally to the moral weight of all conscious experience in ourselves and in all sentient beings, both now and in the future. It is to shake ourselves awake to our evolutionary biases and delusions, and to attempt to see beyond them—even when it is inconvenient or violates human norms.

While of course, society needs norms and rules, it is right that we question them, because as history shows, it is often within these norms that the worst moral atrocities hide in plain sight. Blood feuds, cruel and unusual punishment, public executions, child labor, the subjugation of women, and legal racial segregation were the unquestioned norms of previous generations. Conscious Value Realism provides a basis by which to examine our norms, culture, and society, and make urgent changes.

It is important to emphasize that this view merely asserts the existence of a real foundation of morality. The moral landscape to which it points is vast and filled with unknowns. Just as the discovery of the laws of physics opened a universe of further questions, so too does the recognition of a moral reality invite deep exploration into the nature of value, the relationships between different conscious experiences, and the trade-offs inherent to complex moral decisions. There are countless questions yet to be answered: How do we measure and compare different states of consciousness? What is the nature of moral value in extreme or ambiguous cases? How should we weigh the interests of beings vastly different from ourselves? These are not trivial challenges, but they are questions that can only be meaningfully addressed once we agree on the basic truths that underpin them.

We are not isolated islands of awareness, but expressions of a single, ongoing process—a vast unfolding we call "the universe." In a cascade of chemical replication, consciousness was discovered and recruited for its advantages to survival. But as we, consciousness, become aware of what we are, we can begin to influence this evolutionary process, to steer it not merely towards survival, but in pursuit of conscious value itself. We are each of us threads in a seamless tapestry of experience, woven across time and space. To understand this is to know that every act of compassion, every effort to alleviate suffering or nurture wellbeing, is not merely a duty to abstract principles, but a fulfillment of our true nature, of what is real and of what is intrinsically significant.

The possible future of life and consciousness is vast beyond imagining. As human beings, we have a sweeping and deliberative influence over countless trillions of actual and potential sentient beings. As such, we are clothed in both tremendous power and moral responsibility, to avoid unfathomable harms and to chart the vast dimension of mental potential, to become the stewards of consciousness itself.

THE END

APPENDIX

SOLVING THE NON-IDENTITY PROBLEM

Among his many contributions to moral philosophy, Derek Parfit also discovered what is now called the "Non-Identity Problem."[1] The problem arises when an action affects the personal identity of future people in such a way that, had we acted differently, entirely different people would have come into existence. This raises a dilemma: how can we say an action harms future people if, without it, they wouldn't exist at all? Though seemingly abstract, this problem lies at the heart of many of the most pressing moral questions of our time: from climate change to AI, from systemic injustice to animal welfare.

Imagine a policymaker debating whether to implement pollution regulations. Without them, air quality worsens, and future generations suffer more health problems. But the consequences go beyond health: the absence of regulation alters an intricate web of events, changing when people meet, form relationships, and start families, ultimately determining which individuals come into existence.

Every choice, every delay, and every seemingly trivial event sets off a chain reaction that decides which sperm meets which egg, selecting one possible individual from among the trillions of potential genetic combinations that could have emerged in their place. With each variation in history, an entirely different set of people will eventually populate the Earth.

In the case of the pollution policy, the people who would suffer in the degraded environment are an entirely different set of individuals from those who would have existed if the anti-pollution regulations had been enacted. They may have lower-quality lives than people in the less polluted world, but crucially, these specific individuals would not have been born into that better scenario. This raises the question: can we say they have been wronged when their very existence depends on the "bad" decision?

Traditional ethical theories often struggle with the Non-Identity Problem because they ground moral concern in the wellbeing or rights of specific individuals. This is particularly true of what are called person-affecting views, moral views which hold that an action is wrong only if it harms a particular person. Such theories, especially duty-based or rights-based frameworks, are deeply entangled with notions of individual identity. They tie moral obligations to identifiable persons whose interests can be clearly compared across possible outcomes.

For these kinds of theories, moral obligations depend on whether an action makes a particular person better or worse off compared to some alternative scenario in which that same person exists. But when different actions lead to different people existing, there is no one who ends up worse off than they would have been. For all such views, once identity shifts, the basis for moral comparison disappears, thereby creating the Non-Identity Problem.

Conscious Value Realism (which from this point I will abbreviate to "CVR") offers a potential way to resolve this problem. Unlike person-affecting views, it does not ground moral significance in the identities of specific individuals. Instead, it locates moral significance in the intrinsic value and disvalue of conscious states themselves. In this view, what matters is not whether particular individuals are harmed compared to some alternative reality in which they do not exist, but whether the overall space of conscious experience contains more suffering or more flourishing as a result of our actions.

Valenced experiences are not morally significant because of who experiences them, but because they are what make things better or worse in the most basic terms. By directing our attention to these intrinsically valuable and disvaluable states, CVR opens a way to think clearly about moral questions even when no specific person is harmed. If a polluted world produces significantly more suffering than an alternative world in which different people exist but experience greater wellbeing, then all things considered, a less polluted world is morally preferable. The fact that entirely different people exist in each scenario is irrelevant because, under CVR as a foundation, personal identity itself is not the basis of moral concern.

Solving the Non-Identity Problem would have real-world consequences, especially in areas where moral responsibility extends beyond currently existing individuals. Climate ethics is one of the most immediate domains where resolving the Non-Identity Problem matters. The decisions we make today about carbon emissions, deforestation, and environmental degradation will shape the world in which future conscious beings live. And yet the Non-Identity Problem also reaches into a bewildering range of real-world moral issues: reproductive ethics, animal welfare, existential risk, artificial minds, and proposed reparations for slavery. We will explore how CVR might shed light on these thorny moral issues, yet before we do it's worth pausing to consider a legitimate concern that may arise along the way.

Some might worry that this view risks treating individuals as mere vessels or containers for value and disvalue, reducing moral concern to abstract considerations about conscious states rather than respecting persons as unique, irreplaceable beings. This concern echoes longstanding critiques of utilitarianism, particularly the idea that such a view risks overlooking the moral significance of personal identity, autonomy, and the deep connections that people share with one another.

We can respond to this concern in several ways. One response is that CVR does not deny the importance of persons or individuality, but it shifts the basis of moral significance away from the idea that personal identity is a fixed, unchanging essence that makes someone the same person over time. Instead, it focuses on the intrinsic qualities of experience itself—the real, felt realities of suffering and wellbeing—as the foundation of moral concern. The idea that persons are special, irreplaceable entities stems from an intuitive but, arguably, mistaken understanding of the self as a kind of "metaphysical unit" that persists over time. In other words, we often imagine the self as a kind of enduring 'essence,' something that remains fundamentally the same despite changes in our thoughts, memories, and physical form.

As Parfit and others have shown, personal identity is less about some deep enduring essence and more about patterns of psychological continuity. CVR deepens this insight by identifying what truly matters: not the persistence of particular individuals, but the preservation and promotion of intrinsically valuable conscious states. This need not imply that persons are unimportant. Persons, understood as centers of experience, are deeply relevant because they are the bearers of conscious value.

Recognizing the intrinsic value of conscious states does not mean we should disregard the contexts in which they arise. Indeed, promoting flourishing and reducing suffering often requires a deep understanding of the unique ways in which individuals experience their lives, including their sense of autonomy, personal development, and connections with others.

We can also respond to the criticism that CVR undervalues individuals as mere vessels for value by emphasizing that the view does not treat individuals as interchangeable, even if it ultimately grounds morality in experience rather than personal identity. If an individual is suffering intensely, the fact that there is another individual elsewhere who is experiencing happiness does not, under CVR, 'cancel out' the suffering in any meaningful way. Each experience retains its own moral weight; the solution to suffering is not more happiness elsewhere, because under CVR each conscious state discloses its own value or disvalue. It matters in and of itself, not in comparison.

One might further argue that CVR, as an ethical foundation, offers a richer and more respectful account of moral concern than views that treat individuals as distinct metaphysical units. If personal identity is ultimately a construct, then treating it as foundational may obscure a deeper reality: that conscious states are what ultimately matter morally. CVR does not rely on any particular view about personal identity, but it sits comfortably alongside the idea, developed in Chapter 18, that consciousness is a more coherent basis for our identity, because both views shift our concerns about identity to concerns about conscious experience. In this way, CVR does not reduce people to vessels of value; rather, it helps explain why people matter in the first place.

A grounding in CVR does not place personal identity at the center of moral concern, nor does it ignore the fact that many distinct conscious beings have complex and unique inner lives, future-directed aspirations, and deep interpersonal connections. The loss of a loved one is not merely the loss of a unit of value but the disruption of a vast web of meaningful experiences. Rather than dismissing individuality, CVR helps us understand why individuals matter: not because of some enduring metaphysical essence, but because they are the place where conscious value arises. By shifting moral focus from identity to experience, CVR does not weaken our ethical commitments; it deepens them. It provides a way to take moral reality seriously, not by clinging to unstable notions of the self, but by grounding concern in the felt, undeniable significance of conscious states themselves.

With these responses given, let us consider how CVR might shed light on other moral issues that are plagued by the Non-Identity Problem. One example is reproductive ethics. Many ethical concerns about bringing children into existence, such as whether it is wrong to conceive a child into severe poverty or with a debilitating genetic condition, are complicated by the fact that, had they not been conceived in exactly those circumstances, they would not exist at all.

CVR invites us to approach these questions differently. It recognizes that suffering is intrinsically disvaluable, regardless of whether it is experienced by a specific, identifiable individual or by any other conscious being. If a particular reproductive choice leads to more suffering than the alternative, it is morally worse, not because it harms a specific individual, but because it results in a worse distribution of conscious states in the world. This approach allows us to make clear, rational ethical evaluations about whether it is justifiable to bring a child into extremely adverse conditions or whether reproductive choices should be made with the goal of minimizing unnecessary suffering.

One context in which this becomes particularly delicate is disability. Here, it is crucial to distinguish between the moral evaluation of suffering and the value of the persons who may experience it. CVR does not lead us to judge people with disabilities as less valuable, but rather it urges us to recognize that suffering and preventable obstacles to flourishing are disvaluable wherever they occur. A life involving disability can still be profoundly worthwhile, but it may also involve forms of suffering or constraint that no one would rationally choose. A world in which such obstacles are reduced is a better world, not because certain people are less worthy, but because the overall quality of conscious experience is higher.

Many moral debates surrounding future generations revolve around whether we have obligations to people who do not yet exist. For example, should we take measures now to ensure that future societies have access to resources, live in political stability, and are protected from catastrophic risks like nuclear war, pandemics, or artificial intelligence misalignment? The Non-Identity Problem makes this question difficult because any intervention we undertake today will invariably produce a butterfly effect, directly influencing who will exist in the future.

Through a CVR lens, our obligations to future lives are brought into focus. If the future will contain vastly more happiness or suffering depending on our choices, we have a clear moral reason to make choices that will produce a better outcome. This applies not only to preventing specific harms, such as global catastrophes, but also to actively working toward a future in which conscious beings flourish. In this light, CVR offers a stable foundation for longtermist ethics, where shaping the future trajectory of conscious experience takes center stage. By shifting the focus away from personal identity and onto the intrinsic value and disvalue of conscious states, it provides a way to make sense of our moral obligations in areas where traditional person-affecting views often struggle.

One such area is the ethics of systemic injustice and historical wrongs. Traditional moral frameworks often have difficulty explaining why it is imperative to rectify injustices that do not directly harm identifiable individuals. Consider issues like reparations for slavery or colonialism. One common argument against reparations is that no present-day individuals were directly harmed by these historical atrocities: those who suffered under slavery, for example, are no longer alive, and those living today would not exist at all had history unfolded differently. The Non-Identity Problem suggests that, in a certain sense, many descendants of past injustices cannot be considered worse off than they would have been otherwise, since their very existence depends on those past events.

CVR clarifies that what ultimately matters is not who experiences suffering or deprivation, but the fact that such experiences exist. The lingering effects of historical injustices, including economic disparities, systemic discrimination, and generational trauma, contribute to ongoing suffering in conscious experience today. The obligation to address these harms does not rest on the notion of compensating particular individuals who were wronged, but rather on the recognition that these forms of suffering are real, disvaluable, and morally urgent. This reframing moves beyond debates over personal identity and instead focuses on the moral imperative to reduce unnecessary suffering and promote conscious flourishing.

Animal ethics and the treatment of non-human sentient beings also become clearer under CVR's reframing of the Non-Identity Problem. A common argument against prioritizing animal welfare is that the vast majority of farmed animals would not exist at all if not for factory farming. If we ended factory farming, billions of animals that would otherwise be born would never come into existence. From a traditional person-affecting perspective, this raises a moral ambiguity: if the alternative to existence is nonexistence, it seems that we cannot say that the animals are "harmed" by being brought into a world of suffering. The Non-Identity Problem makes it difficult to argue that factory farming is wrong in person-affecting terms because the animals who suffer are only alive due to the very system that exploits them.

CVR offers a way through this ambiguity. It does not matter whether these animals would have existed in alternative scenarios; what matters is that the suffering they experience is real, disvaluable, and morally urgent. The suffering of a factory-farmed pig is just as ethically significant as the like suffering of any other sentient being, regardless of whether that pig would have existed under different conditions. CVR allows us to recognize the moral atrocity of factory farming and similar practices without getting caught in debates about whether it is "better" for animals to exist in suffering rather than to not exist at all.

It is worth pausing here to clarify what this view does and does not imply. None of this is to suggest that we ought to dispense with rights, rules, or virtues. These frameworks have developed over centuries precisely because they reliably support human flourishing and help avoid suffering in many familiar contexts. They offer durable guidance: rules and principles that tend to serve us well in the messy, unpredictable terrain of everyday life, where we cannot see every consequence in advance and where the people involved are usually familiar and continuous over time.

But there are also cases where these familiar frameworks lose their grip: where identity is fragmented, where the agents involved are radically unfamiliar, or where our choices help determine who will exist at all. In these contexts, rules and rights alone often fall short. This is where CVR provides clarity. By grounding moral concern in the structure of experience itself, not in who the experiencer is, we can navigate situations that traditional approaches cannot easily resolve.

One example is the moral status of future digital or posthuman consciousness. If human civilization transitions into a future where minds exist in radically different forms, whether as uploaded minds, artificial minds, or other post-biological entities, questions of personal identity may become deeply ambiguous. For example, if someone uploads their mind to a digital system, is the resulting consciousness "the same" person? If a mind is copied multiple times, do all copies have equal moral status? Traditional ethical frameworks that rely on continuity of personal identity struggle to address such questions.

Instead of asking whether a future digital being is "the same person" as a biological human, CVR is primarily concerned with the nature of its conscious experience. If it can suffer, then its suffering matters. If it can experience joy, then that joy has real moral value. The ethical challenge is not about tracking identities over time, but about ensuring that in whatever forms consciousness might emerge in the future, these are lives that are worth living. This has deep implications for how we design future AI, posthuman societies, and protect conscious entities that do not fit our traditional categories of personal identity.

Resolving the Non-Identity Problem in this way also affects our moral thinking about long-term existential risks. Many existential risk discussions focus on preventing the extinction of humanity, with the assumption that ensuring humanity's survival is a fundamental good. But why exactly should the survival of human beings matter? The Non-Identity Problem complicates this assumption, because if humanity were to go extinct and some other conscious civilization emerged in the future, it would contain entirely different individuals. From a person-affecting standpoint, it is unclear why we should prioritize our specific descendants over other possible conscious beings.

But this overlooks the fact that human beings, unlike most known moral patients, possess a uniquely powerful capacity to influence the future trajectory of conscious experience. Their intelligence, technological abilities, and reflective moral awareness place them in a position of immense instrumental significance. From the perspective of CVR, their continued existence matters not because they are irreplaceable individuals, but because they are among the only known agents capable of either causing or preventing astronomical suffering, and of potentially realizing vast amounts of conscious flourishing.

From a CVR perspective, the moral significance of humanity lies not in our identity, but in our instrumental role as moral custodians: we possess the capacity to shape the future of countless sentient beings, both biological and artificial. Our extinction would not merely mark the end of one species, but could eliminate a uniquely powerful vehicle for moral progress—or worse, clear the way for a future filled with unmanaged suffering. The ethical imperative, then, is not to preserve human identity for its own sake, but to safeguard the value space of conscious experience itself, including the agents most capable of nurturing it.

CVR, ultimately, is a claim about what truly matters. It does not offer a complete taxonomy of valuable states, nor a definitive method for weighing competing experiences. Substantial practical challenges remain in identifying, measuring, and comparing the moral weight of different conscious states—challenges that will likely require ongoing philosophical inquiry, empirical research, and contemplative insight.

What CVR provides is the foundation for asking those questions: grounding moral reasoning in the real structure of conscious experience, rather than in intuitions, conventions, or metaphysical assumptions. In this way, it does not prescribe a single framework for deciding what actions are right or wrong, such as total utilitarianism (which aims to maximize the sum of happiness and minimize the sum of suffering), negative utilitarianism (which emphasizes the urgent priority of preventing suffering over creating happiness), or prioritarianism (which gives extra moral weight to improving the lives of the worst-off). Instead, it clarifies why any of these frameworks matter in the first place: because they attempt, in different ways, to track and respond to the actual moral significance inherent in conscious experience itself.

Importantly, CVR assumes that there really is a fact of the matter about these further questions. Different ethical approaches are not merely expressions of preference or social convention, but attempts to uncover and describe an underlying reality. Some ways of reasoning about morality will be more accurate reflections of this reality than others. By anchoring morality in the undeniable realities of suffering and joy, CVR provides a shared foundation from which competing theories can be evaluated, refined, and tested, so that over time, our understanding of what we should do can grow clearer and more reliable.

In its framing of the Non-Identity Problem, CVR illuminates a range of moral realities that were previously obscured. It clarifies why we should take moral responsibility for shaping the future, even when no specific individuals are harmed by our choices. It removes the conceptual barriers that make issues like climate change, population ethics, artificial intelligence, and historical injustice so difficult to navigate. It provides a way to think about ethics that does not depend on tracking personal identity, instead focusing on the real, undeniable realities of suffering and wellbeing. In doing so, it opens the door to a more coherent and far-reaching moral philosophy, one capable of addressing the most pressing ethical challenges of our time as well as those which lie ahead.

The intuitions that give rise to the Non-Identity Problem are not themselves baseless. They reflect the deep-seated ways we think about persons, continuity, and harm. CVR does not deny these concerns, but it reinterprets them within a framework where moral weight is anchored in experience rather than identity.

Ultimately, CVR does not so much "solve" the Non-Identity Problem as it reframes it, showing that the seeming paradox arises from treating personal identity as the basis of moral concern, instead of the deeper reality of consciousness.

REFERENCES

PART I. THE REALITY OF CONSCIOUSNESS

CHAPTER 1. REAL VALUE

1. Hedda Hassel Mørch, "Is Consciousness a Fundamental Feature of the Universe?", Nautilus, 2017.
2. René Descartes, *Meditations on First Philosophy,* 1641.
3. Daniel Dennett, *Consciousness Explained*, 1991.
4. Sam Harris, *Waking Up: A Guide to Spirituality Without Religion*, 2014.
5. David Chalmers, "Facing Up to the Problem of Consciousness," *Journal of Consciousness Studies,* 1995.
6. David Pearce, The Hedonistic Imperative, online manifesto, 1995.
7. Magnus Vinding, *Suffering-Focused Ethics: Defense and Implications*, 2020
8. Thomas Nagel, *The View from Nowhere*, 1986.
9. Sam Harris, *The Moral Landscape*, 2010.
10. de Lazari-Radek, Katarzyna, and Peter Singer. *The point of view of the universe: Sidgwick and contemporary ethics.* 2014.
11. Sharon Hewitt Rawlette, *The Feeling of Value*, 2020.

CHAPTER 2. MORAL OBJECTIVITY

1. David Hume, *A Treatise of Human Nature*, 1739.
2. Yuval Noah Harari, *The Politics of Consciousness | video lecture*, YouTube, 2021.

CHAPTER 3. INTRINSIC AND INSTRUMENTAL VALUE

1. Sharon Hewitt Rawlette, *The Feeling of Value*, 2020.
2. David Pearce, The Hedonistic Imperative, online manifesto, 1995.

CHAPTER 4. HEDONIC VALUE AND IMPARTIALITY

1. David Pearce, The Hedonistic Imperative, online manifesto, 1995.
2. Susan Wolf, "Meaning in Life," Proceedings of the Aristotelian Society, 2007.
3. Thomas Nagel, *The View from Nowhere*, 1986.
4. Immanuel Kant, Groundwork of the Metaphysics of Morals, 1785.
5. Toby Ord, *The Precipice: Existential Risk and the Future of Humanity*, 2020.
6. Yuval Noah Harari, in conversation with Sam Harris, Making Sense podcast, Episode 71: "Reality and the Imagination," September 8, 2017.
7. Sharon Hewitt Rawlette, *The Feeling of Value*, 2020.

CHAPTER 5. BEYOND THE HUMAN LENS

1. Andrés Gómez-Emilsson, concept of the "qualiascope" as discussed in various essays and interviews on the Qualia Research Institute website.
2. Sharon Hewitt Rawlette, *The Feeling of Value*, 2020.

CHAPTER 6. A NEW REALITY

1. Peter Singer, *Famine, Affluence, and Morality*, 1972.

CHAPTER 7. COSMIC REACH

1. Anders Sandberg, "The Degenerate Era and the End of Time," Future of Humanity Institute, University of Oxford.
2. Toby Ord, *The Precipice: Existential Risk and the Future of Humanity*, 2020.
3. Will MacAskill, *What We Owe the Future*, Basic Books, 2022.
4. Hans Moravec, *Mind Children: The Future of Robot and Human Intelligence*, 1988.

PART II. KNOWN UNKNOWNS

CHAPTER 8. INTO THE UNKNOWN

1. The Cambridge Declaration on Consciousness, 2012, drafted by Philip Low and co-signed by prominent neuroscientists including David Edelman and Christof Koch.
2. Jonathan Birch et al., Review of the Evidence of Sentience in Cephalopod Molluscs and Decapod Crustaceans, LSE, commissioned by the UK government, 2021.

CHAPTER 9. ANIMAL CONSCIOUSNESS

1. Jeremy Bentham, *An Introduction to the Principles of Morals and Legislation* (1789).
2. Peter Singer, *Animal Liberation: A New Ethics for Our Treatment of Animals*, 1975.
3. Sentience Institute, “Global farmed & factory-farmed animals estimates” 2019
4. William MacAskill, *What We Owe the Future*, 2022
5. Richard Dawkins, “But can they suffer?”, Richard Dawkins on Vivisection (2011).

CHAPTER 10. ARTIFICIAL MINDS

1. John Searle, "Minds, Brains, and Programs", *Behavioral and Brain Sciences*, (1980): 417-424.
2. David Pearce, “Humans and Intelligent Machines: Co-evolution, Fusion or Replacement?”, Hedweb (online essay, 2012)
3. Hans Moravec, *Mind Children: The Future of Robot and Human Intelligence*, 1988.

CHAPTER 11. VALUE ASYMMETRY

1. David Pearce, *The Hedonistic Imperative*, 1995 (online manifesto)
2. Toby Ord, *The Precipice: Existential Risk and the Future of Humanity*, 2020.
3. Magnus Vinding, *Suffering-Focused Ethics: Defense and Implications*, 2020.

CHAPTER 12. ALIEN MINDS

1. Enrico Fermi (informal origin), Carl Sagan & Iosif Shklovsky, 1963: "Where is everybody?"
2. Luitfried von Salvini-Plawen and Ernst Mayr, "On the Evolution of Photoreceptors and Eyes", *Evolutionary Biology* 10, 1977.

PART III. BEYOND UTILITY

CHAPTER 13. MORAL PROGRESS

1. Jeremy Bentham, *An Introduction to the Principles of Morals and Legislation*, 1789.
2. John Stuart Mill, *Utilitarianism*, 1861.
3. Sharon Hewitt Rawlette, *The Feeling of Value: Moral Realism Grounded in Phenomenal Consciousness*, 2020.

CHAPTER 14. THE ORGAN DONOR

1. Jeremy Bentham, *An Introduction to the Principles of Morals and Legislation*, 1789
2. John Stuart Mill, *Utilitarianism*, 1861
3. Internet Encyclopedia of Philosophy, "Act and Rule Utilitarianism", last accessed June 2025

CHAPTER 15. THE EXPERIENCE MACHINE

1. Robert Nozick, *Anarchy, State, and Utopia*, 1974

CHAPTER 16. THEORY X

1. Derek Parfit, *Reasons and Persons*, 1984.

CHAPTER 17. LIBERATING CONSCIOUSNESS

1. Stanford Encyclopedia of Philosophy, "Epiphenomenalism," 2024.
2. University of Texas (Austin), *University of Texas Tower Shooting*—autopsy revealing brain tumor pressing against amygdala, August 1, 1966

CHAPTER 18. THE IDENTITY OF CONSCIOUSNESS

1. Derek Parfit, *Reasons and Persons*, 1984
2. Thomas W. Clark, "Death, Nothingness, and Subjectivity," 1994
3. Magnus Vinding, *You Are Them: On Identity, Consciousness, and Ethical Implications*, 2017
4. Daniel Kolak, *I am you: The metaphysical foundations for global ethics.* 2004.
5. Arnold Zuboff, *One Self: The Logic of Experience.* 1990.
6. Rupert Spira, *Being Aware of Being Aware.* 2017.

APPENDIX. CONSCIOUS VALUE REALISM AND THE NON-IDENTITY PROBLEM

1. Derek Parfit, *Reasons and Persons*, 1984.

Made in the USA
Monee, IL
05 November 2026

67815676R00115